Industrial Explosives and their Applications for Rock Excavation

Industrial Explosives and their Applications for Rock Excavation focuses on applications of industrial explosives in civil and mining engineering works. Explosives and their actions are explained in terms of basics, principles, and related chemistry. Explosives and initiation devices are described, including their characteristics, geometry, and timing aspects of the blast design. Designing blasts for rock excavation works is explained, including devices for obtaining large-sized blocks, construction of yards, and excavation of big foundations. Finally, criteria for the mitigation of the associated seismic disturbances are summarized. The book:

- provides an updated vision of industrial explosives, including the best technical advice for rock excavation;
- contains harmonized preliminary modules aimed at introducing basic concepts of chemistry and physics applied to the drilling and blasting technique;
- defines balanced mix of theory capable of providing skills to design an efficient blasting;
- covers excavation problems from different points of view and in different contexts; and
- addresses issues of drilling and loading blast-holes.

Industrial Explosives and their Applications for Rock Excavation is aimed at graduate students, researchers, and professionals in mining engineering and explosives technology.

Industrial Explosives and their Applications for Rock Excavation

Marilena Cardu, Daniele Martinelli, and Carmine Todaro

CRC Press
Taylor & Francis Group
Boca Raton London New York

CRC Press is an imprint of the
Taylor & Francis Group, an **Informa** business
A BALKEMA BOOK

Designed cover image: Marilena Cardu

First published 2024
by CRC Press/Balkema
4 Park Square, Milton Park, Abingdon, Oxon, OX14 4RN

and by CRC Press/Balkema
2385 NW Executive Center Drive, Suite 320, Boca Raton FL 33431

CRC Press/Balkema is an imprint of the Taylor & Francis Group, an informa business

British Library Cataloguing-in-Publication Data
A catalogue record for this book is available from the British Library

ISBN: 978-1-032-14964-6 (hbk)
ISBN: 978-1-032-14968-4 (pbk)
ISBN: 978-1-003-24197-3 (ebk)

DOI: 10.1201/9781003241973

Typeset in Times New Roman
by Apex CoVantage, LLC

Contents

About the authors

Marilena Cardu, Mining engineer, Ph.D., is currently Associate Professor at Politecnico di Torino (Environment, Land and Infrastructures Department – DIATI); she is responsible for the Geomechanics and Geotechnology laboratory in the same Department. Her actual occupational field is education, teaching, and research. The teaching activities include Excavation Engineering and Mining Plants, Underground Works and Mining, and Planning of Sustainable Mining at Politecnico di Torino; Safety and Civil Protection at Tashkent Turin Polytechnic University (TTPU), Tashkent, Uzbekistan; Demolition Techniques, Rock Blasting and control of Unwanted Effects, Civil and Environmental Engineering Ph.D. Main topics of research: mechanical characterization, at different scales (both geometrical and time-dependent), of rocks; rock–tool interaction problems; simulation and optimization of fragmentation, mechanical cutting, rock breaking; evaluation of explosives and initiation systems; industrial explosives for civil demolitions, rock excavation and rock splitting; exploitation of dimension stones; equipment selection in quarrying; controlled blasting: pre-splitting and smooth-blasting optimization; remediation with explosive; vibrations and stresses induced by blasting; environmental impacts due to the exploitation and/or civil demolition (fly-rocks, dust, noise, etc.); organization and planning of excavation techniques.

Daniele Martinelli, Environmental engineer, PhD, is currently Assistant Professor at Politecnico di Torino (Environment, Land and Infrastructures Department – DIATI); he is an expert in numerical modeling and in rock masses characterization for applications both open pit and underground. His teaching activities include: Excavation Engineering and Mining Plants, Planning of Sustainable Mining, MSc in Petroleum and Mining Engineering – Path Mining, Politecnico di Torino; Tunneling and Occupational Safety Engineering, MSc in Environmental Engineering – Path Geoengineering. Main topics of research: Conventional and mechanized excavation, rock mass conditioning for Earth Pressure Balance TBM applications, especially regarding the geotechnical characterization of the conditioned material, Numerical modeling and geomechanics applied to underground mining and civil voids and slope stability.

Carmine Todaro, Environmental engineer, PhD, is currently Assistant Professor at Politecnico di Torino (Environment, Land and Infrastructures Department – DIATI); he is expert in mechanized tunneling, soil and ground improvements, and underground works. The teaching activities include: Excavation Engineering and Mining Plants, Underground Works and Mining, Tunneling and Ground Improvement Engineering at Politecnico di Torino. Main topics of research: Conventional and mechanized excavation, rock mass conditioning for Earth Pressure Balance TBM applications, wear phenomenon of tools in mechanized tunneling, geotechnical characterization of the conditioned material, two-component grout backfilling, soil and rock improvements, and surface and underground mining.

Chapter 1

General aspects of excavation work

1.1 Introduction

Excavation is the process of unearthing soil, rock, or other materials with tools, equipment, or explosives. Coring and drilling activities also fall under this general definition. Moreover, trenching, wall shafts, tunneling, and underground construction are included in the excavation framework. It can be stated that excavation is the preliminary activity of a building project.

In the following, reference is often made to generic rock or soil as a medium.

Excavation works differ from each other according to not only the construction site but also many other characteristics. An attempt at their classification is based on three different characteristics:

The medium where the excavation takes place:

It can range from very hard rock to loose sand, to name the two extremes of a very large scale of materials. For this classification, cohesion can be considered a reference parameter.

The environment:

Excavation works can be open cast, underground, or underwater (dredging).

The purpose:

Reference is made to the excavated material used (quarries and/or mines) or the cavity obtained (civil works).

This classification leads to 12 types (at least) of excavation works and certainly provides a very vague description.

Further important differentiations include

- the cyclical or continuous nature of the excavation;
- the need (or not) to access the site;
- the concomitance (or not) with the construction of the supports.

As mentioned, the purpose of an excavation work is defined taking into account the excavated material used, the cavity produced, or both. The final purpose can affect how work is done.

Examples

1) A trench must be excavated in a given rock. Drill and blast technique can be used, expressly by designing the drilling pattern in such a way as to obtain a muck that can be easily loaded on

DOI: 10.1201/9781003241973-1

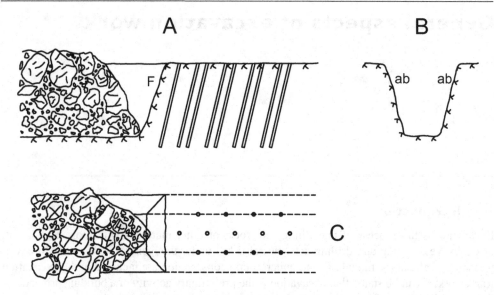

Figure 1.1 Example of excavation of a rock trench, with explosives, for a civil application. The goal is to create a cavity with a predetermined geometry. A: longitudinal section; B: vertical cross section (where ab is the final wall); C: plan view.

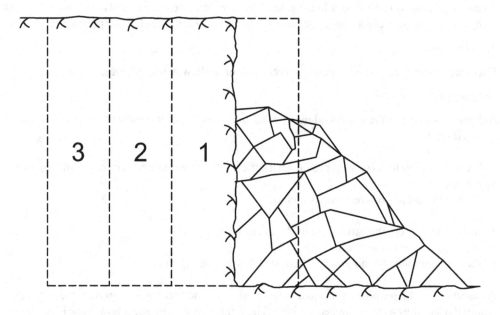

Figure 1.2 Longitudinal section, orthogonal to the excavation face, of an open-pit excavation working with explosives in a crushed stone quarry. The purpose is exploitation to obtain a certain amount of material of a predetermined size. The numbers (1–3) indicate the succession of elementary volumes that must be blasted by groups of mines.

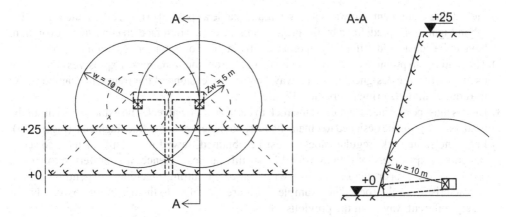

Figure 1.3 Example of arrangement of charges to obtain a certain amount of rock in large, shapeless blocks intended for the construction of maritime work. The purpose is exploitation, similar to that of the case of Figure 1.1b, but with a request for metric or decametric size. Left: plan view; Right: vertical section orthogonal to the face according to the plane A-A.

Figure 1.4 Example of an open-pit construction site for the extraction of granite blocks with explosives. The goal is to obtain regular and healthy blocks from the excavation, rather than shapeless (as in the case of Figure 1.3), which requires very different ways of using the explosive: very small charges, distributed in a large number of parallel and coplanar holes delimiting the blocks to be detached.

the hauling equipment and walls as regular and stable as possible (Figure 1.1); the final walls (ab in Figure 1.1), predefined by the project, and the excavation face, in continuous evolution, have to be considered differently in order to obtain the most appropriate/suitable geometry.

2) The material to produce crushed stone is obtained from the same rock. Explosives can still be used, yet they are designed in such a way that the dimensions of the fragments can be easily introduced into the primary crusher (Figure 1.2).

3) Blocks must be obtained from the same rock to build an artificial reef. The explosives will still be used, but they will be designed to obtain the least possible amount of fine fragments (Figure 1.3).

4) From the same rock, regular blocks must be obtained, sawed into slabs, or used for construction. Explosives are often unsuitable for this purpose. Hence, it is preferred to isolate blocks of the proper shape and size with dedicated cutting machines, depending on the characteristics of the rock. The example in Figure 1.4 refers to the use of explosives, but in a very different way from the previous cases.

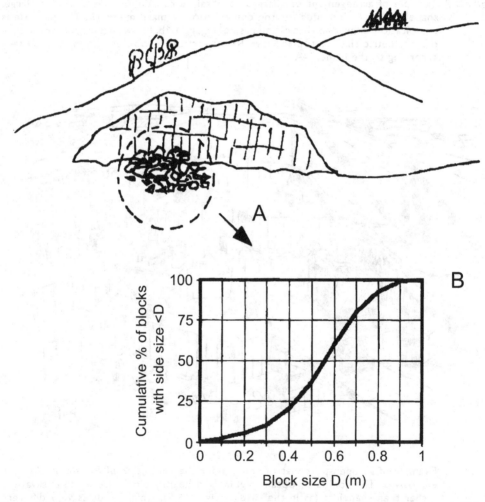

Figure 1.5 Representation of the size of a blasted rock pile (A) through the grain size distribution (B).

Size is defined as the dimension of elements obtained from the excavation. Sometimes, only the maximum size, maximum weight, or volume not to be exceeded (e.g., maximum size side 1.2 m, or maximum weight 5 t, or again maximum volume 1 m^3) can be specified. In other cases, it is necessary to specify the wanted percentages of elements of different sizes (e.g., 10% between 1 and 4 m^3, 30% between 0.5 and 1 m^3, and not more than 20% of debris less than 0.05 m^3), that is, to specify the particle size of the blasted material (Figure 1.5).

The medium in which the excavation develops naturally has a great influence on the excavation work. As previously introduced, the medium can range from loose sand to compact rock.

Only a brief list of the characteristics of the medium to be kept in mind will be given; not all influencing characteristics are quantitatively definable.

Since excavation involves the breaking of the medium, characteristics relating to the behavior of the medium under mechanical stress should be considered for the excavation project in most cases.

Roughly, the following classes can be recognized:

- incoherent media (e.g., loose sand, whose removal is opposed only by weight and friction);
- coherent media with plastic behavior, that is, that undergo progressive strain under stress before failure occurs (e.g., clayey soil, where the bucket of a shovel can penetrate and remove the portion of material that belongs to its size);
- coherent materials with fragile behavior, that is, elastic behavior up to failure, which resist with minimal strain to the applied stresses until they reach a certain level, beyond which they crush (e.g., a solid rock, which must be reduced to fragments from the high pressures resulting from a blast and, therefore, transformed locally into an incoherent medium, which will be removed).

There are, of course, also all intermediate types of behavior, and moreover, the assignment to a category depends on the scale at which the phenomenon is observed: a layer of sand, gravel, and clay can be a consistent plastic medium for an excavator bucket, but individual pebbles are considered differently by the tool of a probe; a whole geological formation can be a plastic medium for orogenic thrusts but not for a dozer blade, and so on.

1.2 Elastic characteristics and mechanical strength of the medium

The mechanical properties of the rocks have a crucial role in the planning of rock excavation and construction for optimum utilization of earth resources with greater safety and least damage to the surroundings. The design and erection of the structure are influenced by the physical-mechanical properties of the rock mass. Young's modulus provides insight into the magnitude and characteristics of the rock mass deformation insinuated by alterations in the stress field (Singh et al., 2012). The elastic characteristics are more observable in rocks with fragile behavior (elastic, or almost, until breakage is reached) and are expressed by the ratio of the applied stresses to the resulting strains (in an elastic medium, these ratios almost do not vary with the load). Most rocks are characterized by elastic properties of an order of magnitude ranging between 10% and 50% of those of steel.

The mechanical strength can be expressed by the stress levels that must be exceeded for failure to occur. It depends on the type of stress (in general, rocks bear tensile stresses from 10 to 20 times lower and shear stresses from 5 to 10 times lower than their compressive strength).

Usually, reference is made to the uniaxial compressive strength (Kahraman, 2001; Chang et al., 2006; Cardu et al., 2012).

It is also noteworthy to evaluate the crushing work, which, for a generic body, is the product of the force that regulates the breakage and the strain tolerated by the body before breaking.

The specific crushing work (Unland and Szczelina, 2004), that is, the one referred to as a unit volume of rock, is approximately proportional to the square of the uniaxial compressive strength and to the first power of the deformability (to be understood as the ratio of the strain to the applied stress). The three above-mentioned characteristics, that is, deformability, uniaxial compressive strength, and specific crushing work, are assessed in the form of pressures. It is useful to check, evaluate, and confirm their orders of magnitude.

1.3 Fractures, cavities, discontinuities, and localized weakening in the medium

When the structure is significantly larger than the rock blocks formed by the discontinuities, the rock mass may be simply treated as an equivalent continuum for the analysis (Brady and Brown, 1985; Brown, 1993; Hoek et al., 1995; Zhang, 2005). Different empirical correlations have been proposed for estimating the properties of jointed rock masses based on the classification indices, such as:

- rock quality designation (RQD) (Deere, 1964; Coon and Merritt, 1970; Serafim and Pereira, 1983; Zhang and Einstein, 2004; Zhang, 2010);
- rock mass rating (RMR) (Bieniawski, 1973; Serafim and Pereira, 1983; Yudhbir and Prinzl, 1983; Nicholson and Bieniawski, 1990; Mitri et al., 1994; Sheorey, 1997; Aydan and Dalgic, 1998);
- Q-system (Q) (Barton et al., 1980; Barton, 2002);
- geological strength index (GSI) (Hoek and Brown, 1997; Gokceoglu et al., 2003; Hoek and Diederichs, 2006).

The favored orientation, the mutual (average) distance between fractures, and so on can be detected in various ways (Hudson and Priest, 1983), but it is difficult to derive an index that allows precise estimates of excavation difficulties. In any case, valid indications for statistical forecasts can be obtained using the reported classification indices.

1.3.1 Rock quality designation (RQD)

Deere (1964) proposed a quantitative index based on a modified core recovery procedure for only sound pieces of a core that are 100 mm or greater in length.

The characterization of rock masses and the evaluation of their mechanical properties are important and challenging tasks in rock mechanics and rock engineering. In many cases, rock quality designation is the only rock mass classification index available (Zhang, 2016).

The method allows using a number to roughly classify the rock mass according to its integrity (Hagan, 1983; Holmberg, 1979). The number, expressed as a percentage, is determined by the frequency of fractures observable in a rock sample taken from the rock mass by coring. It can vary from 0 to 1 (Table 1.1), with 1 (or 100%) being equivalent to a sample with spacing between the fractures above a certain conventional limit. There are also other indices suitable for this purpose, but the RQD is among the most frequently used, as it can be almost immediately determined (Figure 1.6).

Table 1.1 RQD values as a function of rock quality.

RQD (%)	Rock quality
<25	Very poor
25–50	Poor
50–75	Fair
75–90	Good
90–100	Great

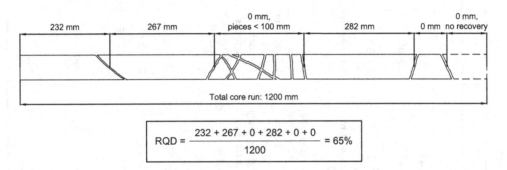

Figure 1.6 The analysis for RQD measurements and calculation. The RQD is defined as the ratio of the sum of the lengths of the elements longer than 10 cm to the total length.

1.3.2   Rock mass rating (RMR) system

The classification of rock masses was proposed by Bieniawski (1973) and can be applied to various rock engineering projects such as tunnels, caverns, slopes, and foundations in civil engineering, as well as haulages and chambers in mining. The classification is based on six parameters: the uniaxial compressive strength of the rock material, the rock quality designation (RQD), spacing, orientation, condition of discontinuities, and groundwater conditions. Each parameter is assigned a given rate to obtain the total rock mass rating (RMR) by summing all these numerical values. The final value is then used for classifying the rock in one of the five rock mass classes proposed by Bieniawski (1979).

For the application of the RMR system, the rock mass is divided into a certain number of "structural areas" so that the different parameters are more or less homogeneous within each area. The limits of these areas often coincide with interruptions in the rock mass's geological and/or mechanical continuity (faults, veins, etc.). The geotechnical classification of the rocks is carried out by applying the data shown in Table 1.2, where the five main parameters are assigned indices or scores (ratings) divided into five intervals. The meaning of the parameters is reflected by the different values of the indices: higher values show better conditions for the rock.

Once the indices have been set, the scores of the five parameters are added together to derive the basic RMR for the area under the exam. The next step is to include a sixth parameter, which takes into account a correction based on the influence of the direction and immersion of the discontinuities, by calibrating the basic RMR using parameter 6 from Table 1.2.

This step is to be treated separately, as the influence of discontinuities depends on the type of work (tunnel, slope, foundation, etc.). Its value is given in qualitative terms.

Table 1.2 Ranges of values are assigned to each parameter for determining the RMR system.

Parameter	Ranges of values						
1. Strength of intact rock material (MPa)							
Point load strength index (MPa)	>10	4–10	2–4	1–2			
Uniaxial compression strength (MPa)	>250	100–250	50–100	25–50	25–5	1–5	<1
Rating (1)	15	12	7	4	2	1	0
2. RQD (%)	90–100	75–90	50–75	25–50	<25		
Rating (2)	20	17	13	8	3		
3. Spacing of discontinuities	>2 m	0.6–2 m	200–600 mm	60–200 mm	<60 mm		
Rating (3)	20	15	10	8	5		
4. Condition of discontinuities	Contiguous discontinuities	Adjacent discontinuities	Partially open discontinuities	Open discontinuities	Highly open discontinuities		
Separation (opening)	<0.1 mm	0.1–0.5 mm	0.5–2.5 mm	2.5–10 mm	>10 mm		
Persistence (continuity)	<1 m	1–3 m	3–10 m	10–20 m	>20 m		
Roughness	Very rough surfaces	Rough surfaces	Slightly rough surfaces	Smooth surfaces	Slikenside surfaces		
Discontinuities' walls	Unaltered	Slightly altered	Highly altered	Completely altered	Residual soil		
Rating (4)	15	10	7	4	0		
5. Groundwater ratio							
Inflow per 10 m tunnel length (l/min)	0	<10	ott-25	25–125	>125		
Joint water pressure at the major principal stress	0	0–0.1	0.1–0.2	0.2–0.5	>0.5		
General conditions	Completely dry	Moist	Wet	Dripping sledges	Water seepages		
Rating (5)	15	10	7	4	0		
6. Orientation of discontinuities							
Direction and immersion of discontinuities	Very favorable	Favorable	Indifferent	Unfavorable	Very unfavorable		
Rating (6)							
Tunnels	0	–2	–5	–10	–12		
Foundations	0	–2	–7	–15	–25		
Slopes	0	–5	–25	–50	–60		

The RMR value can therefore be calculated as follows:

RMR = [1 + 2 + 3 + 4 + 5] + 6

Table 1.3 indicates the classes to which the different rocks can be assigned based on the RMR value, and Table 1.4 shows the meaning of the rock classes.

1.3.3   Rock mass quality (Q system)

This method was developed by Barton et al. (1974) and provides a numerical evaluation of the rock's quality (based on data obtained from about 1,000 cases of tunnel excavation) using six parameters:

- RQD;
- frequency of discontinuities (discontinuity index Jn);
- roughness of discontinuities (roughness index Jr);
- degree of alteration or filling at the edges of the fractures (index of alteration Ja);
- flow through water (reduction index for hydraulic conditions Jw);
- stress conditions (strength reduction factor SRF).

The parameters are used in the following formula, which expresses the quality of the rock mass:

$$Q = \left(\frac{RQD}{J_n}\right) \cdot \left(\frac{J_r}{J_a}\right) \cdot \left(\frac{J_w}{SRF}\right)$$

The classification of the rock mass according to the Q index is shown in Table 1.5.

With regard to the content of this book, relating to excavation with explosives, it should be noted that these classifications pertain above all to the initial decision: whether to use explosives

Table 1.3 Rock classes determined on the basis of the sum of the numerical coefficients.

RMR	Class	Description
100–81	I	Great
80–61	II	good
60–41	III	Fair
40–21	IV	Poor
<20	V	Very poor

Table 1.4 Meaning of the rock classes according to the RMR system.

Class	I	II	III	IV	V
Average time of self-support capacity	10 years	6 months	1 week	10 hours	30 minutes
Free unsupported span	15 m	8 m	5 m	2.5 m	1 m
Cohesion	>400 kPa	300–400 kPa	200–300 kPa	100–200 kPa	<100 kPa
Frictional angle	>45°	35°–45°	25°–35°	15°–25°	<15°

Table 1.5 Rock mass behavior (in the case of tunnel excavation) as a function of the Q index.

Rock mass quality Q	Rock mass behavior
1,000–400	Great
400–100	Extremely good
100–40	Very good
40–10	Good
10–4	Fair
4–1	Poor
1–0.1	Very poor
0.1–0.01	Extremely poor
0.01–0.001	Worst

or to resort to another technique. In worst- to poor-quality rocks, explosive excavation is generally less productive and more dangerous.

1.4   Density

It is necessary to distinguish between on-site density (i.e., of the rock in its natural state) and the density of the excavated material (i.e., loose fragmented rock). In fact, because of fragmentation, an increase in apparent volume is detected (generally from 30% to 50%), hence a decrease in the apparent density of the muck pile compared to the intact rock on site. Usually, reference is made to the density of the water (equal to 1); the same number expresses the density in t/m^3 or kg/l (kg/dm^3). Most compact rocks have on-site densities between 2.5 and 3 t/m^3 and muck density around 2 t/m^3. The density values of the most common rocks and minerals are shown in most of the technical catalogs.

1.5   Abrasivity

In excavation works, apart from certain exceptions, it is necessary to apply considerable forces to the medium, with high unit stresses, thanks to tools of various types or to explosives. Even in the case of explosives, there are two phases where the rock–metal interaction is unavoidable: while drilling the holes and while clearing the material after blasting.

The predictable wear and tear that characterize tools during an excavation process are mainly a function of the mineralogical composition of the medium (the hardness of the components it is made of) and the composition and shape of the tool. The abrasivity of rock and soil is a factor of significant importance for excavation in tunneling, underground construction, mining, or quarrying. Abrasion can be defined as wearing or tearing off particles and material from the solid surface. The Cerchar abrasivity test is nowadays the most used laboratory method to quantify rock abrasivity. It allows to determine an index, the Cerchar abrasivity index (CAI), which is used to evaluate the wear of excavation equipment in different applications such as mining, tunneling, and drilling. This method was initially developed in the 1970s by the *Laboratoire du Centre d'Etude et Recherches des Charbonnages (CERCHAR) de France* for coal-mining purposes (Jacobs and Hagan, 2009; Alber et al., 2014).

The wear of the tools leads to a progressive decline in the performance of the excavation and/ or drilling machines, due to the increase in the contact surface between the tool and the rock (the destructive action of the tool–rock corresponds to a similar rock–tool reaction), until the pressure exerted by the tool is no longer able to cause the rock to fail (Mucha and Krauze, 2018; Käsling and Thuro, 2010). The amount of metal scratched is proportional to the work spent to break the rock; it depends on many factors, not always assessable: composition, geometry, tool behavior, mineralogical composition, structure, texture, mechanical characteristics of the rock, and the presence of discontinuities. The performance of the tools can also be affected by damage of other nature (Mancini and Cardu, 2001; Mucha, 2019), such as breakage or blockage. The ultimate effect is, of course, of an economic nature: from an industrial point of view, the wear (or the specific consumption of tools) must be considered one of the cost parameters for optimizing the cost–benefit ratio (where the benefits are given by the performance of the excavation equipment): if a lower specific consumption of tools is wanted, lower productivity must be accepted (Bilgin et al., 2016). The negative consequence of this statement is that classifications of rocks are not easily practicable in terms of tool wear.

The hardness of minerals is usually evaluated by assigning them a position in the standard "Mohs scale" (West, 1986; Broz et al., 2006; Whitney et al., 2018). On a broader scale, rock hardness is determined by several ways, among which the most suitable is the Vickers or the Knoop test, conducted by evaluating the obtained ratio of the force applied by a standard diamond indenter pressed on a metal surface to the area of the impression (Van-der Voort and Lucas, 1998; Ben Ghorbal et al., 2017; Hutchings, 2009; Oliver and Pharr, 2004; Giannakopoulos and Zisis, 2011). The Knoop hardness test (HK) is typically applied to determine the rocks' micro-hardness by dividing the force by the projected area of the indentation. Test loads are usually in grams-force (gf), whereas indentation diagonals are in micrometers (μm).

Many common minerals are harder than ordinary steel. The use of tools and materials with remarkable hardness and toughness is advised. The tests can be performed on macroscopic or microscopic scales.

1.6 Geometry of the cavity

The geometry of the cavity, developed during excavation, is also of significant importance. The term "geometry" refers to size and shape. The influence of size is obvious; as for the shape, especially in the case of hard mediums, it is necessary to distinguish between two situations: excavation with a single free surface and excavation with several free surfaces.

An excavation face is a surface that delimits the rock to be excavated in relation to the atmosphere. The face is said to have only one free surface when, as in the case of tunnel or shaft excavation, it is surrounded by negative dihedrals, and two or more free surfaces when one or more of the surrounding dihedrals are positive (Figure 1.7).

The excavation is easier with a greater number of free surfaces (this concept shall be resumed later). When there is only one free surface, other surfaces must be generally created with preparatory operations, that is, by locally removing more or less large volumes of rock, making carvings, etc., to facilitate the excavation work. Operations intended to create new free surfaces to facilitate the excavation are called "attacks"; these are followed by a detachment process that allows the rock mass to be removed from its natural state. These two phases are naturally difficult to distinguish, as they typically overlap in the case of excavation in soft ground or in any type of material when the excavation is performed with continuous equipment.

1.7   Environment where the excavation takes place

It affects the way the work is done. As previously mentioned, the main distinction to be made is between open-pit, underground, and underwater excavations.

The underground excavation is subject to severe limitations in terms of size, power, and mode of operation of the machines and often presents difficult safety problems (Cooper, 1978; Cooper et al., 1980; Choullette et al., 1976; Hickson et al., 1985).

Underwater excavation generally faces the challenge of direct access for operators in terms of difficulty or impossibility. Therefore, the underwater excavation needs special methods and equipment.

The excavation work can cause environmental disturbances and damage to third parties. Besides the problem tied to the stability of the cavity and the surrounding rock mass, usually associated with any alteration of the local topography, other harmful effects should be taken into account: ground vibrations, air blasts, noises, dust dispersion, and fly-rocks.

1.8   "Continuous" or "cyclical" nature of an excavation work

The excavation work can take place continuously (a continuous flow of rock is determined, from its natural location to the transporting machines that convey it away from the face), or it can be discontinuous and cyclical.

In both cases, the work is carried out according to several phases (attack, detachment, and removal). However, in the case of continuous excavation, these phases are difficult to distinguish from each other. A typical example of a cyclical excavation is the removal of a compact

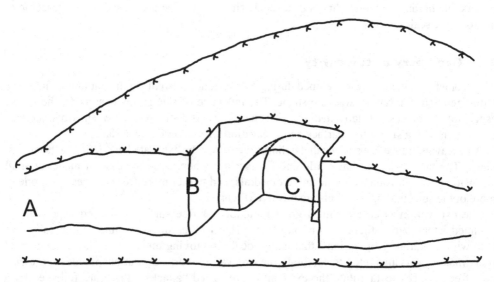

Figure 1.7 An example of the concept of "free surfaces" of the excavation face. The schematic work consists of opening and widening a yard, on a rock slope, and the creation of an attack niche, from which to start a tunnel. In A, there are two free surfaces (face and upper plane); in B, three free surfaces (trihedron); and in C, only one free surface (tunnel face).

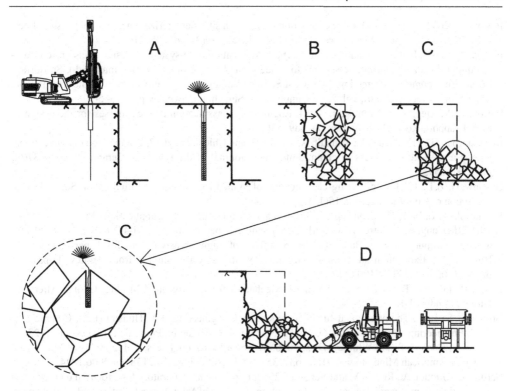

Figure 1.8 Main phases of rock excavation with explosives. A: attack; B: detachment; C: secondary breakage; D: evacuation and haulage to another location.

rock with explosives in a quarry. In this case, the following steps are performed cyclically to guarantee the excavation advancement (Figure 1.8):

A – Attack: the blast holes are prepared in a suitable position, at a certain distance from the face, and the explosive charges are placed.

B – Detachment: charges are initiated; a portion of the rock is broken into fragments and thrown onto the yard.

C – Secondary breakage: too large blocks present in the blasted material are crushed in place with small charges or with other systems.

D – Muck-pile removal (clearing): the resulting rock pile is loaded onto trucks or other vehicles and removed.

References

Alber, M., Yarali, O., Dahl, F., Bruland, A., Käsling, H., Michalakopoulos, N., Cardu, M., Aydin, H. and Ozärslan, A., 2014. ISRM suggested method for determining the abrasivity of rock by the CERCHAR abrasivity test. Rock Mechanics and Rock Engineering, 47, pp. 261–266.

Aydan, Ö. and Dalgic, S., 1998. Prediction of deformation behavior of 3-lanes Bolu tunnels through squeezing rocks of North Anatolian fault zone (NAFZ). Proceedings of the Regional Symposium on Sedimentary Rock Engineering, Taipei, China, pp. 228–233.

Barton, N., 2002. Some new Q-value correlations to assist in site characterization and tunnel design. International Journal of Rock Mechanics and Mining Sciences, 39(2), pp. 185–216.

Barton, N., Loset, F., Lien, R. and Lunde, J., 1980. Application of Q-system in design decisions concerning dimensions and appropriate support for underground installations. M. Bergman (Ed.), Subsurface Space: Environmental Protection, Low-Cost Storage, Energy Savings, Proceedings of the International Symposium (Rockstore'80), vol. 2. Pergamon Press, Stockholm, Sweden, pp. 553–561.

Barton, N.R., Lien, R. and Lunde, J., 1974. Engineering classification of rock masses for the design of tunnel support. Rock Mechanics, 6, pp. 189–236.

Ben Ghorbal, G., Tricoteaux, A., Thuault, A., Louis, G. and Chicot, D., 2017. Comparison of conventional Knoop and Vickers hardness of ceramic materials. Journal of the European Ceramic Society, 37(6), pp. 2531–2535.

Bieniawski, Z.T., 1973. Engineering classification of jointed rock masses. Journal of the South African Institution of Civil Engineering, 15(12), pp. 335–344.

Bieniawski, Z.T., 1979. The geomechanics classification in rock engineering applications. Proceedings of the 4th ISRM Congress, Montreux, Switzerland, September. Paper Number: ISRM-4CONGRESS-1979-117.

Bilgin, N., Copur, H. and Balci, C., 2016. Effect of high abrasivity on TBM performance. In Book: TBM Excavation in Difficult Ground Conditions: Case Studies from Turkey. https://doi.org/10.1002/9783433607190.ch12.

Brady, B.H.G. and Brown, E.T., 1985. Rock Mechanics for Underground Mining. George Allen and Unwin, London, UK.

Brown, E.T., 1993. The nature and fundamentals of rock engineering. J.A. Hudson (Ed.), Compressive Rock Engineering – Principle, Practice and Projects, vol. 1. Pergamon Press, Oxford, UK, pp. 1–23.

Broz, M.E., Cook, R.F. and Whitney, D.L., 2006. Microhardness, toughness, and modulus of Mohs scale minerals. American Mineralogist, 91(1), pp. 135–142. https://doi.org/10.2138/am.2006.1844.

Cardu, M., Giraudi, A., Rocca, V. and Verga, F., 2012. Experimental laboratory tests focused on rock characterization for mechanical excavation. International Journal of Mining, Reclamation and Environment, 26(3), pp. 199–216. https://doi.org/10.1080/17480930.2012.712822.

Chang, C., Zoback, M.D. and Khaksar, A., 2006. Empirical relations between rock strength and physical properties in sedimentary rocks. Journal of Petroleum Science & Engineering, 51, pp. 223–237.

Choullette, D., Clark, G.B. and Lehnhoff, T.F., 1976. Fracture stresses are induced by rock splitters. International Journal of Rock Mechanics and Mining Science, 13(10), pp. 281–287.

Coon, R.F. and Merritt, A.H., 1970. Predicting in situ modulus of deformation using rock quality indexes determination of the in situ modulus of deformation of rock. ASTM International, pp. 154–173.

Cooper, G.A., 1978. Method and Apparatus for Breaking Hard Compact Material Such as Rock. U.S. Patent 4,099,784.

Cooper, G.A., Berlie, J. and Merminod, A., 1980. A novel concept for a rock breaking machine II: Excavation techniques and experiments at large scale. Journal of the Royal Society of London, 873(1754), pp. 352–372.

Deere, D.U., 1964. Technical description of rock cores for engineering purposes. Rock Mechanics and Engineering Geology, 1, pp. 17–22.

Giannakopoulos, A.E. and Zisis, T., 2011. Analysis of Knoop indentation. International Journal of Solids and Structures, 48(1), pp. 175–190.

Gokceoglu, C., Sonmez, H. and Kayabasi, A., 2003. Predicting the deformation moduli of rock masses. International Journal of Rock Mechanics and Mining Sciences, 40(5), pp. 701–710.

Hagan, T.N., 1983. The influence of controllable blast parameters on fragmentation and mining costs. Proceedings of the First International Symposium on Rock Fragmentation by Blasting, Luleå, Sweden, pp. 31–51.

Hickson, P.W., McMahon, D.W. and Schapper, S., 1985. The shape of hard rock mining in the 21st century. Proceedings of the First International Federation of Automatic Control, IFA C, Symposium, Brisbane, Queensland, Australia, July 9–11.

Hoek, E. and Brown, E.T., 1997. Practical estimates of rock mass strength. International Journal of Rock Mechanics and Mining Sciences, 34(8), pp. 1165–1186.

Hoek, E. and Diederichs, M.S., 2006. Empirical estimation of rock mass modulus. International Journal of Rock Mechanics and Mining Sciences, 43(2), pp. 203–215.

Hoek, E., Kaiser, P.K. and Bawden, W.F., 1995. Support of Underground Excavations in Hard Rock. A.A. Balkema, Rotterdam.

Holmberg, R., 1979. Design of tunnel perimeter blast hole patterns to prevent rock damage. Proceedings of the 2nd International Symposium Tunnelling 79, Institute of Mining and Metallurgy, London, pp. 280–283.

Hudson, J.A. and Priest, S.D., 1983. Discontinuity frequency in rock masses. International Journal of Rock Mechanics and Mining Sciences & Geomechanics Abstracts, 20(2), April 1983, pp. 73–89.

Hutchings, I.M., 2009. The contributions of David Tabor to the science of indentation hardness. Journal of Materials Research, 24(3), pp. 581–589.

Jacobs, N. and Hagan, P., 2009. The effect of stylus hardness and some test parameters on CERCHAR abrasivity index. 43rd Rock Mechanics Symposium and 4th Rock Mechanics Symposium ARMA, Asheville, June 28–July 1, pp. 9–191.

Kahraman, S., 2001. Evaluation of simple methods for assessing the uniaxial compressive strength of rock. International Journal of Rock Mechanics and Mining Sciences, 38(7), October, pp. 981–994.

Käsling, H. and Thuro, K., 2010. Determining abrasivity of rock in the laboratory. Engineering Geology. Technische Univ. München, Germany, pp. 1973–1980.

Mancini, R. and Cardu, M., 2001. Scavo in Roccia-Gli Esplosivi. Hevelius Ed., Benevento (Italy), 297 pp.

Mitri, H.S., Edrissi, R. and Henning, J., 1994. Finite element modeling of cablebolted stopes in hard rock ground mines. Proceedings of the SME Annual Meeting, Albuquerque, USA, pp. 94–116.

Mucha, K., 2019. The new method for assessing rock abrasivity in terms of wear of conical picks. New Trends in Production Engineering, 2(1), pp. 186–194.

Mucha, K. and Krauze, K., 2018. Planning of experiment for laboratory tests on rock abrasivity. Mining-Informatics, Automation and Electrical Engineering, 3, pp. 17–24.

Nicholson, G.A. and Bieniawski, Z.T., 1990. A nonlinear deformation modulus based on rock mass classification. International Journal of Mining and Geological Engineering, 8(3), pp. 181–202.

Oliver, W.C. and Pharr, G.M., 2004. Measurement of hardness and elastic modulus by instrumented indentation: Advances in understanding and refinements to methodology. Journal of Materials Research, 19(1), pp. 3–20.

Serafim, J.L. and Pereira, J.P., 1983. Consideration of the geomechanical classification of Bieniawski. Proceedings of International Symposium on Engineering Geology and Underground Construction, vol. 1. A.A. Balkema, Rotterdam, pp. II.3–II.44.

Sheorey, P.R., 1997. Empirical Rock Failure Criteria. A.A. Balkema, Rotterdam.

Singh, R., Kainthola, A. and Singh, T.N., 2012. Estimation of elastic constant of rocks using an ANFIS approach. Applied Soft Computing, 12(1), January, pp. 40–45.

Unland, G. and Szczelina, P., 2004. Coarse crushing of brittle rocks by compression. International Journal of Mineral Processing, 74(Supplement), December 10, pp. S209–S217.

Vander Voort, G.F. and Lucas, G.M., 1998. Microindentation hardness testing. Advanced Materials & Processes, 154(3).

West, G., 1986. An observation on Mohs' scale of hardness. Quarterly Journal of Engineering Geology and Hydrogeology, 19, pp. 203–205, https://doi.org/10.1144/GSL.QJEG.1986.019.02.12.

Whitney, D.L., Fayon, A.K., Broz, M.E. and Cool, R.F., 2018. Exploring the relationship of scratch resistance, hardness, and other physical properties of minerals using Mohs scale minerals. Journal of Geoscience Education, 55, pp. 56–61.

Yudhbir, W.L. and Prinzl, F., 1983. An empirical failure criterion for rock masses. Proceedings of the 5th International Congress on Rock Mechanics, Melbourne, vol. 1, pp. B1–B8.

Zhang, L., 2005. Engineering Properties of Rocks. Elsevier Ltd., Amsterdam.

Zhang, L., 2010. Estimating the strength of jointed rock masses. Rock Mechanics and Rock Engineering, 43(4), pp. 391–402.

Zhang, L., 2016. Determination and applications of rock quality designation (RQD). Journal of Rock Mechanics and Geotechnical Engineering, 8(3), June, pp. 389–397.

Zhang, L. and Einstein, H.H., 2004. Using RQD to estimate the deformation modulus of rock masses. International Journal of Rock Mechanics and Mining Sciences, 41(2), pp. 337–341.

Chapter 2

Explosions and explosives

2.1 Explosions

The term "explosion" applies to various types of phenomena characterized by a significant release of energy in a very short time span (see Figure 2.1).

Observations and explanations

- In cases A, B, and C, an explosion takes place, as the container is strong enough to sustain itself until failure from peak pressures is reached inside (the explosion is the sudden release of energy that follows the collapse of the container).
- In cases D and E, the explosion takes place regardless of whether the container exists or not (the pressure reached does not depend on its presence or strength).
- In case F, the container collapses, but no explosion occurs, as the pressure applied into the medium is practically incapable of contraction and expansion (e.g., water).
- In cases A, B, and F, only physical phenomena are involved (injection of a pressurized fluid into the container, heating and evaporation of a liquid previously introduced). In the same cases, the energy released upon failure was induced from the outside (as pressurization, as heating).
- In cases C, D, and E, the energy is generated from a chemical reaction that took place inside the container (when it exists): from the outside, only the initiation of the reaction is provided (triggering).

This book deals with cases like types C, D, and E (or rather, with the practical application of phenomena of those types).

Many tests can be used to clarify the concepts of explosion and explosive intuitively and are carried out to verify whether a substance is "explosive", for the purposes of transport regulation (Fujishiro et al., 1974; Green et al., 2020; Suceska, 2012; Vermorel et al., 2017; Berthoud, 2000).

2.2 Explosives

La polvere da sparo altro non è che una materia contenente nei suoi pori aria estremamente condensata. . . . Basta aprire le piccole cavità dove quest'aria condensata è racchiusa ed immediatamente essa esce fuori con la massima forza. – Gunpowder is nothing but a matter containing extremely condensed air in its pores. . . . Just open the small cavities where it is enclosed and immediately it comes out with the maximum strength.

(Euler, 1760)

DOI: 10.1201/9781003241973-2

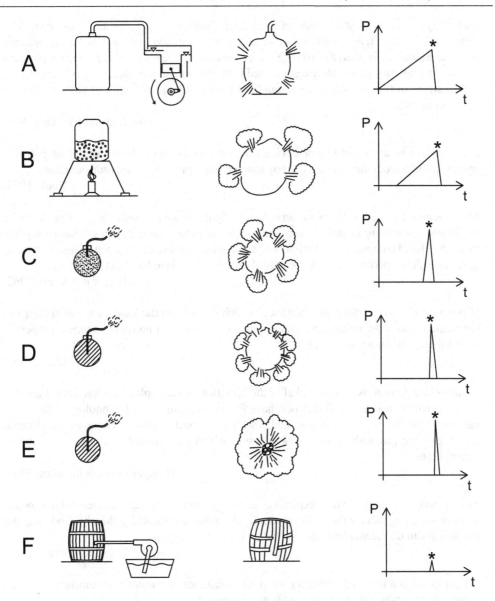

Figure 2.1 Examples of explosion. The graphs on the right project the pressure trend over time for the six cases examined; the asterisks indicate the peak values. A: tank progressively pressurized with compressed air; B: pressure cooker without safety valve; C: bomb loaded with black powder; D: bomb charged with detonating explosive; E: bare charge of a detonating explosive; F: barrel filled with water and pressurized.

Les matières explosives . . . sont des substances qui, sous l'influence de la chaleur ou d'un choc considérable, se décomposent en dégageant un grand volume de gaz à haute température – Explosive materials . . . are substances that, under the influence of the heat or of a considerable shock, decompose by releasing a large volume of gas at high temperature.

(Chalon, 1911)

Unter Explosion versteht manhierbei einen Zerfallsvorgang, bei dem die Ausgangstoffe in sehr rascher chemischer Reaktion in neue Stoffe übergehen, wobei gleichzeitig erhebliche Wärmeund auch grosse Gasmengen frei werden – An explosion is a process of decay where the originary materials are converted into new substances due to a very fast chemical reaction, so that considerable amounts of heat and large amounts of gas are released at the same time.

(Beyling and Drekopf, 1936)

An explosive is any substance or device which produces, upon the release of its potential energy, a sudden outburst of gas, thereby exerting high pressures on its surroundings.

(Cook, 1958)

On désigne sous le nom d'explosif un composé défini, ou un mélange de corps susceptibles de dégager en un temps extrémement court un grand volume de gaz portés à haute température – A defined compound, or a mixture of substances capable of releasing a large volume of gases at high temperature in an extremely short time, is referred to as an explosive.

(Berger and Viard, 1962)

[E]xplosives are, by definition, substances which exhibit similar kinetics under similar circumstances, since the remarkable fact is that there are almost no other chemical properties which they would possess as a class.

(Macek, 1962)

Within some thousands of a second after the initiation of the explosive, there occurs a series of events which, in drama and violence, have few equivalents in civil technology. The chemical energy of the explosive is liberated, and the compact explosive becomes transformed into a glowing gas with an enormous pressure which can amount to and exceed 100,000 atmospheres.

(Langefors and Kihlström, 1967)

From a practical point of view, explosives are simply materials that are intended to produce an explosion, i.e., to have the ability to rapidly decompose chemically, thereby producing hot gas which can do mechanical work on the surrounding material.

(Persson et al., 1994)

An explosive is a material, either a pure single substance or a mixture of substances, which is capable of producing an explosion by its own energy.

(Davis, 2016)

The given definitions are taken from works of different eras (to understand the suggested definition – explanation of Euler – it has to be considered in the frame of 1760 chemistry, during which it was still in its early stages and every gas was qualified as air), but they are quite similar; it is interesting to notice that the archaic definition of Euler is very similar to modern and general ones provided by Cook and Langefors, who suggested that the mechanical effects of the explosive can be explained by the velocity at which the developed gas particles collide against the walls of the charge container ("it comes out with maximum force"; "outburst of gas . . . thereby exerting high pressures"; "and the compact explosive becomes transformed into a glowing gas with an enormous pressure"). The definition of Chalon underlines the need for a powerful trigger, whereas Beyling and Drekopf highlight the exothermicity of the chemical reaction. Conversely, Berger and Viard, Persson et al., and Davis emphasized the notion of explosives belonging to the category of chemical substances or mixtures.

In all practical applications, an explosive is characterized by an explosion reaction that must occur only when the substance undergoes a rather powerful stimulation (heating, shock); this stimulation is called a "trigger". It is also necessary for the triggered reaction to not stop and proceed quickly until the complete decomposition of the charge is reached. An explosive (Definition I) can therefore be defined as

> a substance (compound or mixture) in a condensed status (solid or liquid) which, following a suitable localized stimulation (trigger), reacts exothermically, in a very short time, with final products completely, or largely, gaseous, at the temperature reached by the reaction.

The simplest example, from a chemical point of view, is the liquid oxygen/coal mixture ("oxyliquite", an explosive rarely used today; see Figure 2.2).

The reaction is completed much faster than the gas expansion time: there is a very short time during which the 22.4 NL (normal liters) of carbon dioxide are contained in only 37 cm³, and that's why the pressure, in that very short amount of time, is enormous. The proposed example's reaction is very simple: $C + O_2 = CO_2$.

Observations and explanations

- There are mixtures of gases (e.g., oxygen/hydrogen, oxygen/methane, and similar) or mixtures of gases and solids (e.g., air/sawdust and the like) which, after being exposed to a flame or a spark, react explosively. They are not considered "explosives" because they are not condensed (this makes their use and storage too complicated).

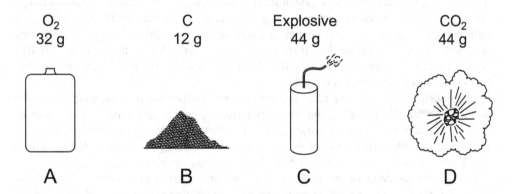

O_2	C	Explosive	CO_2
32 g	12 g	44 g	44 g
A	B	C	D

Figure 2.2 Example of an explosive mixture (oxyliquite). A. Oxidizing agent: liquid oxygen. Mass 32 g (1 gram-molecule). Density: 1.14 g/cm³. Volume: 28 cm³. B. Fuel: carbon. Mass: 12 g (1 gram-molecule). Density: 1.4 g/cm³. Volume: 9 cm³. C. Mixture: oxyliquite. Mass: 44 g. Density: 1.19 g/cm³. Volume: 37 cm³. D. Fumes: carbon dioxide. Mass: 44 g (1 gram-molecule). Normal volume: 22,400 cm³. Density before reaction: 1.19 g/cm³; density during the reaction: 1.19 g/cm³; density after the reaction, under normal conditions: 1.96×10^{-3} g/cm³.

- There are mixtures of solids that, when triggered, react quickly and exothermically, but are not considered explosive, because the final products are not gaseous (e.g., the aluminum/iron oxide mixture used for welding).
- In an explosion reaction, if the final product is water, it is considered gaseous (since the temperature reached is always much higher than the critical water temperature).
- The oxylicite mixture described in Figure 2.2 is prepared by immersing porous carbon cartridges in liquid oxygen; obviously, it deactivates after some time due to oxygen evaporation. The other explosives contain oxidizing oxygen as a component of an easily decomposable substance and therefore have a long life (years).
- Many explosives are also made up of inert substances (which do not take part in the reaction) or substances that react by producing solid final products (solid residues).

2.2.1   Velocity of explosion reactions: types of explosions

A reaction of explosion is defined as a fast chemical reaction. However, two categories of explosion reactions can be recognized: "deflagrations" and "detonations". The formers are characterized by "relatively low" reaction velocities, whereas the latter display fast reaction propagations at very high velocities. In any case, "slow" and "fast" are adjectives; hence, for an engineering approach, a quantitative reference is needed.

The reference velocity is the sonic velocity that propagates into the explosive: the "deflagrations" propagate with a subsonic velocity, while the "detonations" with a supersonic velocity (see Figure 2.3).

Observations and explanations

- Intuitively, deflagration propagates because the developed hot gases from the combustion of one particle reach another and bring it to the reaction temperature, heating it in return. A detonation propagates because a sudden local increase in pressure (and temperature) brings the substance to the reaction conditions. Hence, the energy developed by the reaction is (partially) used to pressurize the adjacent layer of substance, where the phenomenon is repeated and so forth, until the exhaustion of the explosive substance. Since the wave velocity (the pressure variation propagating in a medium is a pressure wave) is the sonic one, the detonation propagates *at least* with sonic velocity.
- As previously stated, there are non-ideal detonations in which the reaction zone is not very thick. In the "ideal" detonation, the "detonation front" shows clear discontinuity in terms of pressure and composition, which moves into the explosive (at the "detonation speed").
- In particular conditions, a translation from detonation to deflagration, from deflagration to detonation, or also, of course, interruptions in the reaction can occur.
- Strictly speaking, deflagration or detonation is not an explosive characteristic. Indeed, the velocity of the explosive reaction depends on how the reaction is triggered and other factors. However, explosives must be distinguished from "deflagrating explosives" (category 1) and "detonating explosives" (category 2) according to the ease in obtaining the type of reaction and their method of employment.
- Some "deflagrating" explosives, with well-controllable reaction speeds, are called "propellants" because they are usually adopted for propulsion purposes (e.g., rockets); they have had some usefulness in peculiar rock blasting works.

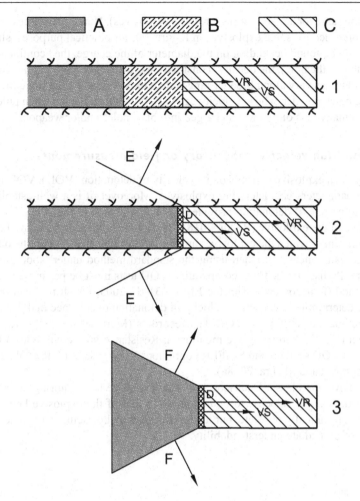

Figure 2.3 "Snapshots" of explosive charges during the reaction. 1. Deflagrating charge confined in a solid medium. 2. Detonating charge confined in a solid medium. 3. Charge
detonating in the atmosphere.

A. Fumes (reaction products). B. Reaction zone. C. Unturned explosive (in cases
2 and 3, also "unaware" of the initiation of the reaction). VR: reaction velocity;
VS: sonic velocity of the explosive; D: detonation front; E: shock front in the solid
medium; F: shock front in the air.

The schematization of the detonation is simplified: the detonation front can differ
considerably from a flat surface, and the reaction zone's thickness can be negligible.
In these cases, the detonation is defined as "non-ideal detonation".

2.2.2   Velocity of explosion reactions: typical values for different explosives

Manufacturers provide the values of the reaction velocities (deflagration or detonation, depending on the type of explosive) for commercial explosives. These values are measured with standard

tests (but not the same for all laboratories or for all explosives). Numerical values considered for reference are provided for some explosives in Figure 2.4; for practical purposes, slightly different velocities can be found depending on the diameter of the charge, the "confinement" (a bare charge, in a tube, in the hole), the density (constipation), the trigger type, and the grain size. The deflagration velocities vary along broader ranges than those of detonation (an extreme example is given by the black powder, whose velocity can vary from about 1 cm/s, when phlegmatized in safety fuses, to many tens of m/s when charged into blast holes and/or weapons).

2.2.3   Detonation velocity: laboratory or field measurements

The efficiency of an explosive depends on its velocity of detonation (VOD). VOD is indeed one of the critical parameters characterizing explosives. Moreover, it is a well-established reality that measuring the rate of detonation provides a good indication of the strength and, in return the performance of the explosive (Reddi, 2018). Its value is influenced by many factors related to the conditions under which the explosives are used. The VOD measured in the blast hole may differ from the value measured when using the standard method under laboratory conditions (Mertuszka and Pytlik, 2019). Thus, comparative VOD tests must be performed, including the start–stop method (in accordance with the EN 13631–14:2003, which is based on a point-to-point method, determining the average velocity of detonation over a specified distance) and the continuous methods, consisting of VOD/data recorders (Mertuszka et al., 2018). Among the continuous methods, the following are mentioned: Resistance wire continuous VOD method, TDR continuous VOD system, and SLIFER continuous VOD system (Tete et al., 2013; Mishra et al., 2019; Agrawal and Mishra, 2018a).

Figure 2.5 shows the classic Dautriche test, a comparative test requiring a detonating cord with a known detonating velocity, to be compared to that of the explosive being tested. The Dautriche test is technically obsolete, but its description may be useful from an academic point of view being of immediate understandability.

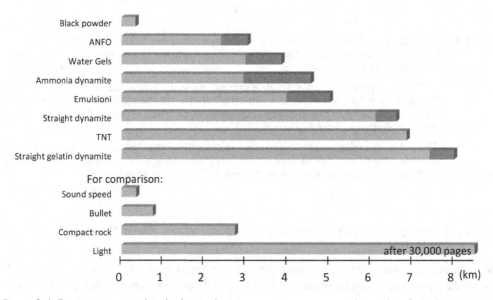

Figure 2.4 Distance covered in 1 s by explosion reactions, compared to other fast phenomena.

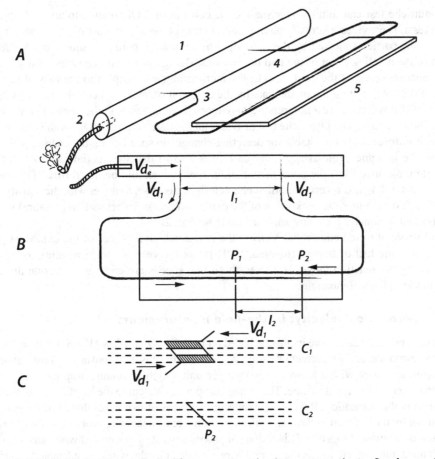

Figure 2.5 Scheme of the Dautriche test for measuring the detonation velocity D_e of an explosive.

    A. Overview of the test device.

        1: test cartridge, on a sand layer;
        2: triggering;
        3, 4: detonating cord branches characterized by a known velocity D_1;
        5: metal plate (lead, soft iron).

    B. Scheme of the measurement. l_1 is the basis of the measurement, on the cartridge. The two branches of the detonating cord depart from its extremities. P_1 is the point, on the plate, where the detonation fronts of the two cords would meet if the detonation started at the same time from the two ends of the base (i.e., if D_e were infinite). P_2 is the point, observable after the test, where the two fronts actually meet: it is marked by an inclined notch on the metal. l_2 is the distance of the meeting point from P_1. With simple developments, it can be written as

$$\frac{l_1}{V_{d_e}} = \frac{2l_2}{V_{d_1}}$$

    and since l_1, l_2, and D_1 are known, D_e can be obtained.

    C. Explanation of the mechanism that causes the indentation: in C_1, immediately before they meet, the (conical) collisions of the two cords travel against each other; in C_2, at point P_2, the instant of the impact, they collide along a generator and the remarkable pressure leaves an oblique notch well observable on the metal.

The Dautriche test can intuitively visualize the concept of VOD related to an explosive subject to tests. It provides a sort of "photograph" that captures the impact of the collision fronts between the particles ejected by the two adjacent detonating cords, as shown by the detailed sketch of the meeting point of the two detonations. Being known and constant (about 7 km/s), the detonation speed of the cord strands, the inclination (with respect to the axis of the metal plate) of the notch left on the metal plate by the collision of the two shock fronts, gives an idea of the speed at which the reaction products are ejected (some km/s, in the immediate vicinity of the exploding body). The Dautriche test provides an average value of VOD of the charge under test but is unfortunately unsuitable for detecting changes in speed along the charge, which may be sensitive in some explosives. In any case, it is known that the maximum VOD stabilizes at a certain distance from the point of initiation. Lastly, it should be noted that VOD (usually given in m/s or km/s) is a very important characteristic because, together with the density of the explosive, it determines the peak value of the explosion pressure (detonation pressure) which is an important parameter for many applications (Shekhar, 2012).

It is important to remember that VOD is the rate at which a "state" of the explosive propagates, that is the fact of being detonated. VOD is not the velocity corresponding to the gases produced by the reaction move. However, these two speeds are correlated, despite the differences in their physical meaning.

2.2.4 Detonation velocity: in-the-hole measurements

The continuous resistance wire method is based on the basic Ohm's law ($V = RI$), where V = voltage, R = resistance, and I = current intensity. When the current is held constant against a shortened (i.e., detonated) wire with a known resistance per unit length, a voltage drop can be measured instantaneously at any point in time. The voltage drop is equivalent to the length of resistance wire consumed in the detonation. Resistance wire probes consist of two wires that must be physically shorted out by the detonation through ionization. Some resistance wire probes consist of just two insulated wires twisted together. Other kinds of probes consist of one coated wire placed inside of a small metal tube which acts as the second wire. Granted that the wires are adequately shortened during the detonation, the resistance wire method does provide actual continuous VOD along the explosive column due to the high sampling rates ranging from 1.25 MHz to over 10 MHz. If the wires are not adequately shortened in a continuous and reliable trend, errors occur, and excessive electronic noise and severe dropouts are commonly obtained (Agrawal and Mishra, 2018b).

Detailed monitoring of hundreds of single-hole and multi-row blasts in full-scale varying blast environments was completed for specific analyses. To always obtain more reliable results, high-speed photography and/or videography, laser surveying, refraction seismographs, digital seismographs to measure ground vibration/airblast, fragmentation and jointing analysis systems, borehole inspection cameras, multi-channel recorders, and continuous VOD instrumentation have been used (Chiappetta, 1998). However, rarely only one of the above-mentioned technologies has been used alone. In fact, most of field tests were equipped with multi-instrumentation systems to generate the most meaningful interpretation of the results for use in blast design optimization (Olmsted et al., 1998; Palangio et al., 1997).

With the apparatus shown in Figure 2.6, the intact length of a coaxial cable placed in a charged blast hole is measured and recorded at intervals of 10 μs. The cable length indicator is given by the reflection time of an electrical pulse. The example refers to a hole charged at the bottom with a faster explosive and for the remaining part with a slower explosive. The method is only applicable if the trigger is placed at the bottom of the hole.

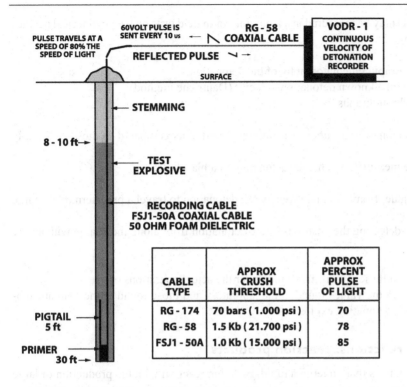

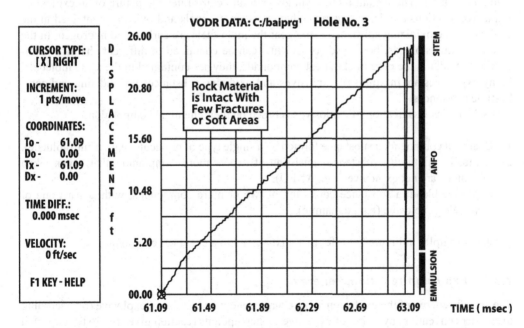

Figure 2.6 Schematic representation of a method for the continuous detection of detonation velocity in a blast hole and example of recording.

Source: Chiappetta (1993)

In synthesis, VOD may be measured in a laboratory or in field tests on bare cylindrical charges or in standard casings:

- by means of ultra-rapid cinematography of the detonation;
- by comparison with a known detonation velocity (Dautriche method);
- with electronic chronographs.

In these tests, the detonation occurs under different conditions compared to those commonly adopted.

VOD can also be measured on charges detonating in a blast hole:

- as the average value, from the time taken by the detonation to travel a predetermined charge length;
- continuously, by detecting the positions of the detonation front along the charge with a very fine time scan.

Indeed, the velocity in these tests is measured under the actual conditions of use.

Differences between the values found in the laboratory and those found in the field are substantial in "not ideal" detonating explosives.

2.3   Explosion reactions: reaction products

Explosives are substances that undergo a rapid oxidation reaction with the production of large amounts of gases. The instant increase in gas pressure constitutes the nature of an explosion. Explosive reactions are decompositions that consume the fuels and oxidants contained in the explosive. The oxidizing substance is oxygen; the main fuels are carbon and hydrogen. In the case of mixtures/blends, both oxidizer and fuel can be contained in different chemical compositions, while in the case of chemical compounds, they are contained in the same molecule. Many explosives are mixtures of explosives and oxidizing substances or oxidizing and combustible substances.

A schematic classification of explosives is based on their chemical composition:

1) Chemical compound: a substance formed by a single type of molecule (e.g., trinitrotoluene).
2) Explosive mixture: a substance consisting of two or more compounds, that, when taken separately, are not explosive (e.g., ANFO).
3) Explosive blend: a substance consisting of two or more compounds, with at least one of them being explosive (e.g., dynamite).

The two examples in Figure 2.7 refer to a mixture and to a chemical compound.

2.3.1   Explosion reactions: energy

The specific energy of the explosion is the energy generated by 1 kg of explosive (or, according to another convention, by 1 liter of explosive), whereupon its reaction gives rise to the expected final products (fumes and solid residues). It can be calculated as the difference between the heat of formation pertaining to elements of the final products and that of the explosive substance (the

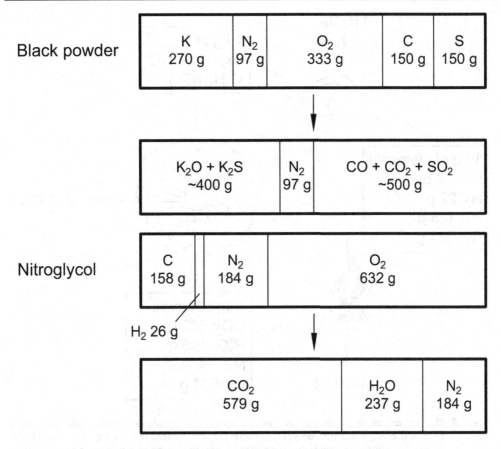

Figure 2.7 Examples of reaction products from an explosive mixture with solid residues (black powder) and a compound without "solid residues" (dinitroglicol). The first deflagrates; the second is a detonating compound.

heat of formation per gram-molecule and/or per gram of the various substances are underlined in the chemistry explosive manuals).

The example proposed in Figure 2.8 explains in a very intuitive way the concept of the specific energy of the explosion. Specifically, the example refers to a chemical compound (trinitroglycerin, initials NG).

Considering now that the same amount of fumes is obtained from 1 kg of explosive through the explosion reaction, the amount of heat developed is equal to the difference between the two mentioned (in this case, 1,467 Cal/kg or 6.14 MJ/kg); this is the specific explosion energy of the tested substance.

The concept illustrated in Figure 2.8 has a general validity: an explosive substance provides an exothermic product, that is, the transformation from original elements to the final ones (fume and solid residues) produces energy. An endothermic product would give rise to more energy in the explosion reaction but spontaneously explode too easily.

A comparison of the specific energies of some explosives is shown in Figure 2.9.

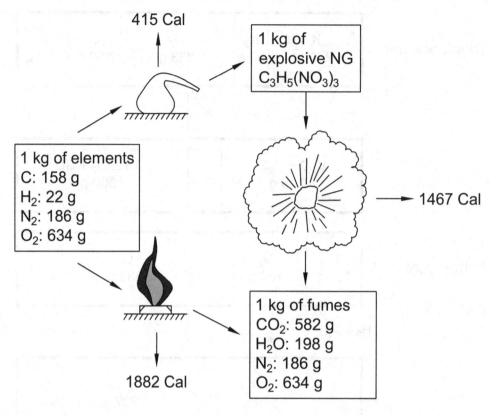

Figure 2.8 Illustration of the concept of specific energy of explosion, with reference to a chemical compound (nitroglycerin, NG). Upper branch: from 1 kg of chemical elements (combustible, oxidizing, and inert elements), 1 kg of explosives and a certain amount of heat are obtained by an exothermic chemical process (415 Cal). Lower branch: from 1 kg of elements, 1 kg of fumes could be obtained with normal combustion and an amount of heat greater than the one provided by the above process (1,882 Cal).

Observations and explanations

In addition to thermochemical calculations, the specific energy of an explosion can be directly measured by bomb-calorimeter tests (Peralta et al., 2001), conceptually similar to the tests of the calorific value of fuels (van Kessel et al., 2004).

A different approach for the specific energy assessment resides in the analysis of effects obtained by the explosion of a charge in a homogeneous medium characterized by known (and constant) physical–mechanical characteristics. For example, lead (Pb) is the base for the "Trauzl test" (Gordon et al., 1955; Ko et al., 2017; Afanasenkov, 2004), while the use of water is foreseen in the "test of the pool explosion" (Satyavratan and Vedam, 1980). The latter can separately provide information regarding energies associated with the impact effect and the expansion of the gases developed by the explosion. Ballistic mortar tests are another approach: the specific energy is deduced by assessing the propulsive effect of a bullet, knowing both the mass of the explosive charge and that of the bullet (Taylor and Morris, 1932; Yoshida et al., 1985; Leiper, 1989; Elshenawy et al., 2019).

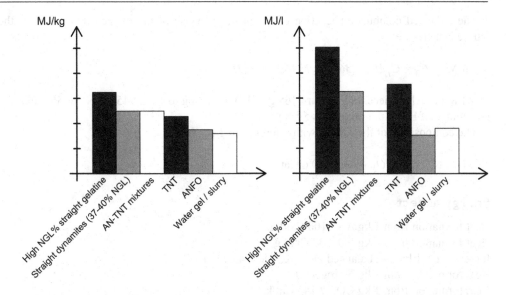

Figure 2.9 Comparison of specific energies referred to the mass unit (weight S.E., MJ/kg) and the volume unit (volumetric S.E., MJ/l) of some explosives. The volumetric S.E. refers to the theoretical loading density, that is the density of the explosive. Note that the ranking of the TNT, which is fourth in terms of weight SE, and second in terms of volumetric S.E., is due to its high density.

All these tests can be defined as "indirect" and are more considerable than calorimetric determinations (theoretical or experimental). Indirect tests are more used for production controls and comparisons between different explosives or batches of the same explosive.

2.3.2 Analysis of an explosion reaction (balanced ANFO)

An explosive is defined as "balanced", or "zero oxygen balance", when the oxygen released from the oxidizing part of the mixture, or molecule, equates to the amount necessary to oxidize all the carbon into CO_2, and all the hydrogen into H_2O (and, if other fuels are present, to oxidize them completely, e.g., aluminum to Al_2O_3).

The following example refers to the ammonium nitrate/fuel oil mixture (ANFO), a most widely used blasting agent.

Ammonium nitrate (NH_4NO_3 – AN, molecular weight 80.05) decomposes according to the reaction:

$$NH_4NO_3 = N_2 + 2H_2O + \tfrac{1}{2} O_2$$

(80.05 g of AN make therefore available 0.5 molecules, that is, 16 g of oxygen).

The fuel oil (FO), for simplicity, assumed to be $C_{12}H_{24}$, molecular weight 168, burns according to the reaction:

$$C_{12}H_{24} + 18O_2 = 12 CO_2 + 12 H_2O$$

(168 g of fuel oil requires 18 molecules, that is, 576 g of oxygen).

The balanced combustion reaction of FO at the expense of the oxygen developed by the nitrate is therefore:

$$36\ NH_4NO_3 + C_{12}H_{24} = 36\ N_2 + 12\ CO_2 + 64\ H_2O$$

(2881.8 g of AN is necessary to burn 168 g of FO; referring to 1 kg, 945 g AN per 55 g FO; the percentage of FO in the mixture is 5.5%).

The composition of the fumes, with reference to 1 kg, is:

331 g N, 173 g CO_2, 496 g H_2O (steam).

Energy aspect

Heat formation from 1 kg AN:1100 Cal/kg.
Heat formation from 1 kg FO: 235 Cal/kg
It means, for 1 kg of a balanced mixture, 1,052.4 Cal/kg
Heat formation from 1 kg Nitrogen: 0
Heat formation from 1 kg CO_2: 2,143 Cal/kg
Heat formation from 1 kg $H_2O_{vap.}$: 3,209 Cal/kg
Then, for 1 kg of fumes, 1,962.4 Cal/kg
The explosion-specific energy is: 1,962.4 − 1,052.4 = 910 Cal/kg (3.81 MJ/kg)

Observations and explanations

From the reaction equations for ANFO, one can readily see the relationship between oxygen balance, detonation products, and heat release. The equations assume an ideal detonation reaction, which in turn assumes a complete mixing of ingredients, proper particle sizing, adequate confinement, charge diameter and priming, and protection from water.

Fuel oil is a variable mixture of hydrocarbons and is not precisely CH_2, but this identification simplifies the equations and is accurate enough for the purposes of the following examples: in reviewing these equations, keep in mind that the amount of heat produced is a measure of the energy released:

$$(94.5\%\ AN) - (5.5\%\ FO): 3NH_4NO_3 + CH_2 \rightarrow 7H_2O + CO_2 + 3N_2 + 0.93\ kcal/g \qquad (1)$$
$$(92.0\%\ AN) - (8.0\%\ FO): 2NH_4NO_3 + CH_2 \rightarrow 5H_2O + CO + 2N_2 + 0.81\ kcal/g \qquad (2)$$
$$(96.6\%\ AN) - (3.4\%\ FO): 5NH_4NO_3 + CH_2 \rightarrow 11H_2O + CO_2 + 4N_2 + 2NO + 0.60\ kcal/g \quad (3)$$

Equation (1) represents the reaction of an oxygen-balanced mixture containing 94.5% AN and 5.5% FO. None of the detonation gases are poisonous, and 0.93 kcal of heat is released for each gram of ANFO detonated. Equation (2) showcases a mixture of 92.0% AN and 8.0% FO, where the excess fuel created an oxygen deficiency. As a result, the carbon in the fuel oil is oxidized only to CO, a poisonous gas, rather than the relatively harmless CO_2. Because of the lower heat of the formation of CO, only 0.81 kcal of heat is released for each gram of ANFO detonated. In equation (3), the mixture of 96.6% AN and 3.4% FO has a fuel shortage, creating an excess oxygen condition. Some of the nitrogen from the ammonium nitrate combines with this excess oxygen to form NO, which will react with oxygen in the atmosphere to form extremely toxic NO_2. The heat absorbed by the formation of NO reduces the heat of the reaction to only 0.60 kcal, which is considerably lower than that of an over-fueled mixture. Also, the CO produced by an over-fueled mixture is less toxic than NO and NO_2. A slight oxygen deficiency is preferable (Dick et al., 1986).

Observations and explanations

A "balanced" explosive makes the most of the fuel in its composition from an energetic point of view (Rowland and Mainiero, 2000; Mainiero, 1997).

Almost all explosives for non-military uses are balanced or not far from balance.

Furthermore, a balanced explosive provides completely oxidized, and non-toxic, end products (CO_2, water, nitrogen, and metal oxides).

An excess of oxygen above zero balance may be required to virtually exclude CO production for explosives to be used underground.

However, some oxygen-deficient explosives can also be used because they excel in certain characteristics, such as high energy/volume ratio (instead of energy/mass), high detonation rate, safe handling, and insensitivity to thermal stimuli or shock. A typical example is $TNT - C_7H_5 (NO_2)_3$, where oxygen is not even enough to oxidize all the carbon in its molecule to CO. Characteristic data of TNT: detonation velocity 6.8 km/s, density 1,600 kg/m³, specific energy 4.31 MJ/kg (6.89 MJ/l); Final products: CO, methane, hydrogen, and unburnt carbon. TNT is a military explosive but is also used in industrial explosives, mixed with oxidizers to balance it. The most used oxidizer is NH_4NO_3; the typical mixture is the balanced *amatol* (about 20 parts by weight of TNT and 80 parts of NH_4NO_3). Characteristic data of a balanced Amatol: detonation velocity $\cong$ 5.4 km/s, density $\cong$ 1200 kg/m³, specific energy $\cong$ 4.6 MJ/kg (5.2 MJ/l); final products: CO_2, N_2, and water.

2.4 Energy and power

Often reference is made to the "power" of an explosive or a charge to express characteristics not related to the actual power (work/time relationships), but specific energies or pressures. The real "power" developed by a charge (relationship between the energy developed and the duration of the reaction) is enormous and clearly differentiates the explosives, as well as the explosions from ordinary combustion. The example in Figure 2.10 refers to three explosives with different reaction rates; these reactions are compared with the common combustion, where the same volume of substance develops much more energy, but much slower.

The thermal power is computed as the ratio between the total energy and the reaction duration.

For case D, different scenarios can be speculated in function of the flow rate of the fuel oil mixture:

$Q = 1$ l/min: duration 11,760 s, thermal power: 664 kW.
$Q = 10$ l/min: duration 1,176 s, thermal power: 6.64 MW.
$Q = 100$ l/min: duration 117.6 s, thermal power: 66.4 MW.
$Q = 1000$ l/min: duration 11.76 s, thermal power: 664 MW.
The black powder is reached at 319,000 l/min.
The dynamite is reached at 11,580,000 l/min.

2.5 Specific work of the explosion

The value obtained by converting into mechanical units the difference between the heat of formation of the fumes and that one of the explosives represents only an "indicator" of the work

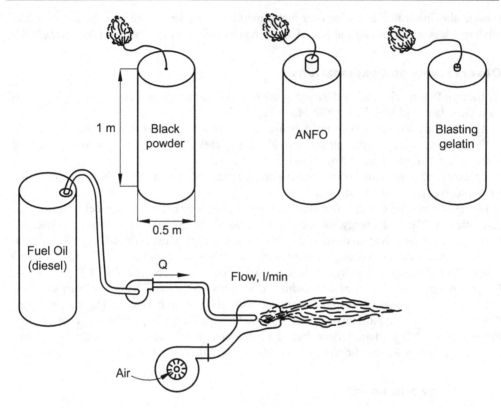

Figure 2.10 Power (MW) developed by explosive charges of the same volume and shape, compared to the thermal power obtainable from an equal volume of ordinary fuel, at various consumption rates.

Example	Reference volume (L)	Mass (kg)	Specific energy (MJ/kg)	Total energy (MJ)	Reaction duration (s)	Thermal power (MW)
A	196	235	3	705	1/300	212
B	196	176	3.81	669	1/2,500	1,675
C	196	284	4.5	1,278	1/6,000	7,698
D	196	186	42	7,812	Function of the flow rate	Function of the flow rate

Note: In the case of fuel, the oxidizer (air) is to be added separately.

that can be obtained from a unit mass of explosive. Similarly, the calorific value of a fuel is an indicator of the work that can be achieved by burning 1 kg of it in an engine. An alternative way of understanding the work potentially obtained from the explosion can be explained by the system described in Figure 2.11.

2.6   Concluding remarks on the concept of "specific energy"

The use of the specific energy concept implies the intention to consider and treat the explosive as a thermal machine. It is sometimes immediately applicable (e.g., in the case of launch explosives);

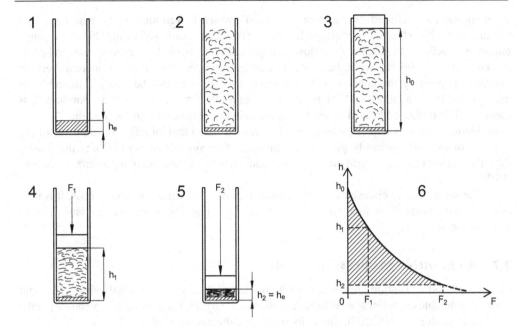

Figure 2.11 Work obtainable from an explosion (considering the explosion equivalent to a thermal machine).

1: 1 kg of explosive is placed in an indestructible and insulated cylindrical tank with a unitary cross surface. For simplicity, the volume of the explosive is indicated by h_e.

2: The reaction takes place; at the end of the reaction, solid residues and fumes are observed at atmospheric pressure in the tank; up to this point, no work has been obtained or dispersed; the gases are hot, but no heat has been dispersed.

3: A piston is placed on the fumes. The volume occupied by solid residues and fumes is equal to h_0.

4, 5: The piston is pushed down; the force required F_1 is initially small (equivalent to an intermediate volume h_1) and gradually grows, as the diagram indicates; the test stops when all the fumes in the volume originally occupied by the charge have been compressed, that is, when the final volume reached h_2 is equal to h_e. In this condition, the value of F_2 is the maximum. The volume is occupied by a mixture of hot gases and solid residues.

6: The area included in the dashed triangle is the work done to return to the initial volume, and it is also the work that could have been obtained from the explosion.

but in other cases, especially in rock blasting, it is only possible with many precautions. The main difficulties are caused by the following factors:

- the vagueness of the expansion ratio (differently from a blast, both the volume of the combustion chamber and the displacement are known in an engine, whereas in a blast hole only the former is known; consequently, the "efficiency" can be only speculated);
- the application of rules valid for relatively slow phenomena to very rapid phenomena: the inertia plays in fact a crucial role in blasting;
- the uncertainty related to "performance" and "efficiency", two concepts necessary for defining, from an engineering point of view, the "useful effect" of an explosion.

In an engine, the useful effect is a "work rendered", which is well measurable with tests; in a blasting work, it is a "result", certainly a function of the "work rendered" by the "blast" machine, but not well defined. In fact, for an explosion, a qualitative score (well succeeded, fair, and poor) can be estimated after the work, but these adjectives cannot be translated into incontrovertible numbers. Explaining the concept with a numerical example cannot be easily concurred, for example having available "4 MJ, 1 is spent in useful effect and 3 are wasted". Anyhow, this aspect will be further addressed by analyzing the effects of explosives on the medium.

Reflecting on the examples given in § 2.4, it can be stated that an effective indicator of the "power" of an explosive can be given by the product of the specific energy (MJ/kg), the density (kg/m³), and the reaction velocity (m/s): this product has the dimension of a power/surface area (MW/m²).

On the same note, to choose the most suitable explosive for certain specific needs, it is necessary to understand if the final goal can be reached by choosing "work" or "pressure", as an astute parameter.

2.7 An intuitive gas pressure model

Figure 2.12 and its caption are taken from a popular work of Bragg and Williams (1934) and have considerable educational effectiveness (almost all physics texts, in various versions, make use of the "analogy of billiard"). The balls symbolize the gas molecules.

In case of an explosion, it can be imagined that balls are initially stationary and linked together by chemical bonds, forming the molecules of the explosive substance. The explosion reaction is the element that provides the spheres with enough energy for breaking the bonds.

The "pressure" (with reference to Figure 2.12, the thrust on the mobile side) depends on

- the number of balls on the table (density);
- speed at which the balls are moving (temperature).

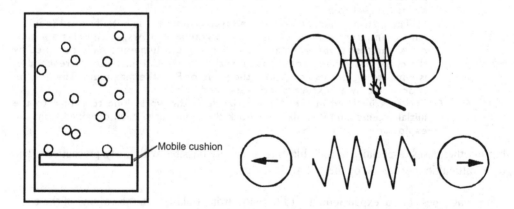

Figure 2.12 Diagram of gas pressure, through the analogy of the billiard table with a mobile side: "Diagram of a small experimental billiard table with a mobile side. The moving balls push the movable side in front of them, losing part of their energy in this work. If, on the other hand, the mobile side is pushed forward suddenly, the motion energy of the balls grows".

Source: Bragg and Williams (1934)

The sketch next to the billiard graphically explains the occurrence of the explosion: the balls (atoms and molecules) are connected to each other by a wire, which holds a compressed spring. With a match, the wire is burned (this phase is the trigger of the reaction); the spring snaps and pushes the balls, which acquire a certain speed (the released energy corresponds to the elastic strain energy stored in the springs).

2.8   Explosion pressure

It is quite obvious that when an explosion occurs, from 1 kg of solid, about 1 Nm^3 (normal cubic meter) of gas is produced; consequently, the pressure of the fumes is very high. However, the term "pressures" is used in the plural. In fact, in the case of detonation, two different effects can be distinguished. The "quasi-static" pressure (Zhang et al., 2018; Wang et al., 2012; Feldgun et al., 2015), also called "chemical pressure" (Zhong and Tian, 2013), is caused by the confinement of fumes in a small volume (and which, once the reaction is complete, persists throughout the volume). The "detonation pressure" or "peak" or "shock" pressure exists only during the detonation and where the reaction takes place (detonation front).

The difference between the "quasi-static pressure" (the meaning of "quasi" will be taken up later) and impulsive pressure can be visualized with a simple test (Figure 2.13).

A unit mass is suspended at a given height on a very accurate dynamometer. The mass is dropped and hits the plate in a very short time (Δt). The dynamometer records, in that very short time, a high load (F_d, dynamic load), but after that, a lower load (F_s, static load) is constantly

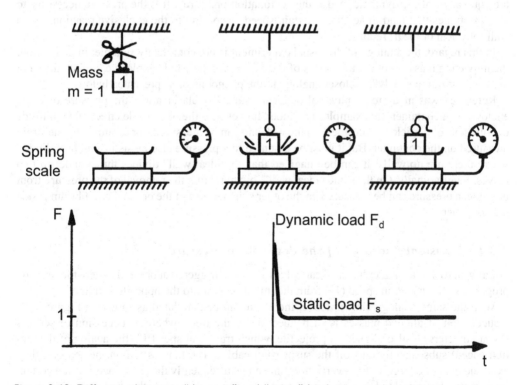

Figure 2.13 Difference between "dynamic" and "static" load generated by the same body.

displayed. The latter is simply the weight of the fallen mass. The dynamic load is the force needed to nullify the speed (Δv) of the mass in the time Δt (duration of the impact):

$$F_{\mathrm{d}} = M\,\frac{\Delta v}{\Delta t}$$

The static load is simply $M \times g$ (g = acceleration of gravity) and lasts until the mass is removed.

Observations and explanations

The conceptual difference between the static and dynamic loads is often not well understood, and spirals back to the well-known puzzle: "Does 1 kg of iron weigh more than 1 kg of feather?". Obviously, they both weigh 1 kg, but, by dropping them on a plate, it can be easily seen that the duration of the impact is much longer for 1 kg of feather than for 1 kg of iron, as the latter is more rigid, and therefore, the balance can go to full scale for when the iron is dropped on it.

The equivalent of the static load is the "quasi-static" pressure of the reaction products on the walls (ideal or real) of the container where the reaction occurs. This pressure cannot be calculated by applying the ideal gas law ($PV = nRT$) or by considering a correction in terms of constant covolume; in the following, an example of calculation with a plausible semi-empirical formula will be examined.

The term "static" (or quasi-static") is used to highlight the equal distribution of the pressure in all the volumes occupied by fumes. With reference to the simple example previously described, the equivalent of the dynamic load is the "detonation pressure". It is the pressure necessary to stop the fumes that tend to acquire a certain speed, according to the explosive reaction, that a wall (physical or ideal) retains.

In this regard, the analogy of the visual experiment is not completely suitable: in detonation, the impacting mass is not the total mass of the explosive: the whole explosive, in fact, does not detonate "simultaneously". A closer analogy to the phenomenon is presented below.

Reference was made to a "physical or ideal wall". For clarification, the pressure of a gas enclosed in a container, for example, a cylinder, however, is the observable effect of a multitude of collisions of particles (molecules) on the wall and can be measured, for example, by the strain measured on the container. But it must be assumed that pressure also exists inside the cylinder, where there are no walls. It can be imagined that an "ideal wall" divides the cylinder in two halves, longitudinally. On the same "ideal wall", by imagining the removal of the pressure from one side, a pressure can be measured on the other side because of the original equilibrium inside the container.

2.8.1 Transient "nature" of the detonation pressure

Usually, in rock blasting, cylindrical cartridges are used, triggered at one end, where the reaction progresses with a certain speed (V_{d}) from the initiation point to the opposite extreme.

With the same analogy shown above, the event can be simulated as shown in Figure 2.14: a succession of aligned masses is suspended above the floor, where a succession of sensors (dynamometers) is also aligned. A projectile, which travels at speed V_{p} (the analog of detonation speed) subsequently cuts off the suspension cables. The masses fall on the sensors; they exert the dynamic force F_{d} for a very short time and subsequently the static load F_{s}. At a certain instant, the dynamic load F_{d} occurs only in one point; in the points that had previously been

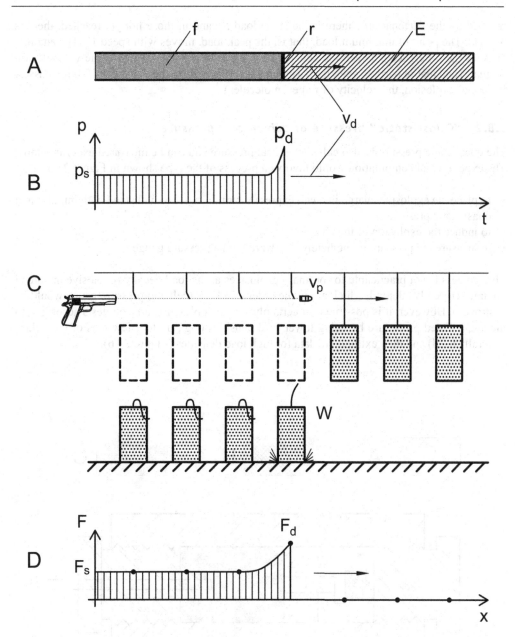

Figure 2.14 Visualization of the detonation pressure concept as the propagation of a dynamic overload.

A. Representation of the explosion of a cylindrical charge (confined) at a given instant *t*. *f*: fumes; *r*: reaction zone; *E*: intact charge; V_d: detonation velocity.

B. Pressure diagram along the axis of the charge at the same instant *t*. p_s: quasi-static pressure; p_d: detonation pressure; V_d: detonation velocity.

C. Mechanical analogy. V_p: velocity of a bullet, which serves as a signal; W: velocity of the falling masses.

D. Load diagram at the instant considered. F_s: static load; F_d: dynamic load.

reached by the "detonation", there is the static load F_s, and in those not yet reached, there is no load. The point of maximum load, that is, the peak load, moves with speed V_d. The analogy clarifies the difference between the concept of detonation velocity (which is the velocity of a "signal") and that of particle velocity W (in this case, W would be the velocity of falling masses; in case of explosion, the velocity of fumes' molecules).

2.8.2 "Quasi-static" pressure or "chemical" pressure

The quasi-static pressure is also called chemical pressure and can be measured experimentally. The experimental computation would involve a process of the type shown in Figure 2.15

- to place the explosive charge in a cell expressly built in order to withstand both impulse and quasi-static pressure;
- to induce the explosion of the charge;
- to measure the pressure immediately afterward with a pressure gauge.

This process is not practicable for ordinary explosives at the load density (explosive mass/cell volume) currently adopted, which are in the order of 1 kg/L (cell and pressure gauge would be destroyed). However, it is possible (but certainly very complex) to carry out tests according to the above-listed procedure by using lower load densities (e.g., up to a few tens of the values generally used) and then extrapolate data for high load densities (Figure 2.16).

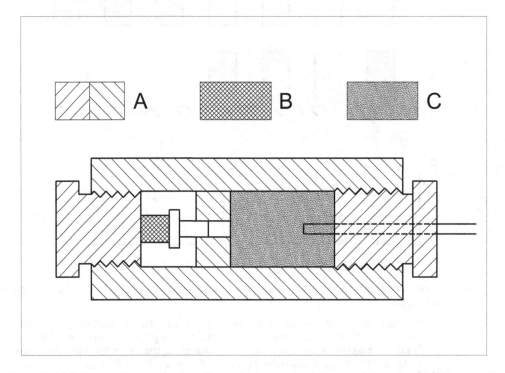

Figure 2.15 Scheme of a device for direct measurement of explosion pressure. A: steel; B: deformable metal (copper); C: explosive. Note that the device cannot be used by completely filling the combustion chamber with explosives.

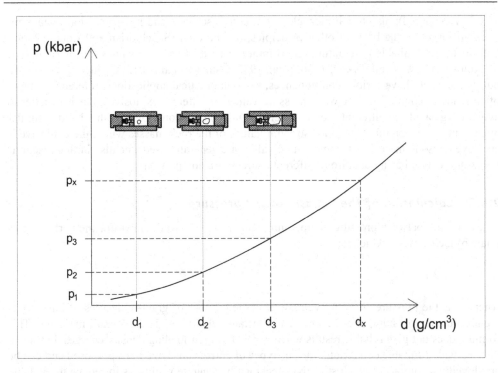

Figure 2.16 Scheme for the evaluation of the explosion pressure by extrapolation from direct tests. On the vertical axis are the pressures; on the horizontal axis are the loading densities.

The apparatus is loaded with increasing charges (therefore, the volume of the cell being constant, with increasing densities), and the tests are carried out in succession, not exceeding the pressure value compatible with the integrity of the apparatus (points d_1p_1, d_2p_2, and d_3p_3). The p_x value at the right density, higher than the measurable limit, is estimated by extrapolation. The equation of the curve is experimentally derived.

In any case, the result is affected by doubts about the validity of the extrapolation. It is therefore better to obtain the value through calculations. The necessary data are the heat of the reaction, the explosion reaction, and the density of the explosive.

Observations and explanations

At loading densities of about 1 g/cm³, pressures can vary from 2 to 3 kbar for black powders to a few tens of kbar for the most common industrial explosives. Lower pressures are of interest in the field of powders used in firearms (where often the pressures in the combustion chamber are of about 100 bar only); in these applications, the loading density is much less than 1 g/cm³.

Very low loading densities (from 0.5 g/cm³ downwards) and pressures lower than 1 kbar are important only for splitting purposes (Bauch and Lempp, 2004; Wang et al., 2014).

The "pressure gauges" used for measuring the pressures of the explosion gases are not conventional: more often, they are "crushers" (copper cylinders, obviously disposable, which are plastically crushed by pressure; the pressure value reached is obtained from the shortening that they have endured); also tough piezoelectric pressure sensors, and high-precision dynamom-eters operating on the basis of other principles, can be used (Sanchidrián and López, 2006). Different texts available in literature report rather discordant pressure values for the same type of explosive (Cowan and Fickett, 1956; Deal, 1957; Östmark et al., 2012; Gogulya et al., 2004), but this does not have serious consequences, since in practical applications, reference is more often made to comparisons between pressure values of different explosives, but it is better to avoid using data from different sources when making comparisons. On the other hand, a similar uncertainty arises on the "dynamic" strength values of the rock masses where the explosive is employed; also in these cases, comparative values are generally used, and also in this case, it is necessary to avoid using data from different sources as a comparison.

2.8.3   Calculation of the "quasi-static" pressure

A relationship between pressure, temperature, and volume is well known for the perfect gases ("law of ideal gases") which is:

$$P V = n R T$$

where P is the pressure, V is the volume, n is the number of gram-molecules of gases, T is the absolute temperature, and R is the "gas constant", that is, 8.315 J/°K gram-molecule. This formula does not give reliable results when applied to high-loading density charges, but it can be used with a suitable correction. Various types of corrections have been proposed and can be employed; the simplest (which still gives technically accurate results) is to reduce the volume that must be considered in the calculation by deducting a certain amount, which is a function of the loading density. It should be noted that this correction is not linked to the covolume correc-tion relative to temperature and pressure values. The corrective term (α) is plotted as a function of the inverse of the loading density in Figure 2.17 (Cook, 1958); the loading density scale has been added to the original graph, for simplicity of use. In practice, a covolume correction is car-ried out by subtracting an (empirical) covolume, which is a function of the loading density. It is assumed that the function is valid for all explosives.

The formula for calculating the pressure becomes:

$$P = n R T/(V - \alpha(V)).$$

2.8.3.1   Example of calculation of the explosion pressure

Reference is made to the ammonium nitrate-fuel oil (ANFO) mixture, used at the loading den-sity of 0.8 kg/L (1.25 L/kg). First, the explosion temperature is evaluated, dividing the heat developed by 1 kg of the mixture by the resulting gas mixture's thermal capacity (at constant volume).

The composition of the fumes and the heat of the explosion are calculated for the theoretical reaction, and the pressure calculation refers to this hypothesis too.

The components of the fumes and their specific heats at constant volume (c_v) are:

CO_2 (17.3 %): $c_v = 0.29 - 87.8/T$ Cal/kg
N_2 (33.1 %): $c_v = 0.234 - 49/T$ Cal/kg
H_2O (49.6 %): $c_v = 0.943 - 1153/T$ Cal/kg

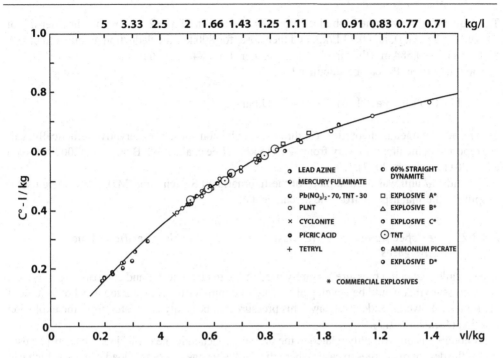

Figure 2.17 Correlation between the covolume and the loading density.

Source: Cook (1958)

From these data, the average specific heat (at constant volume) of the gas mixture is obtained (percentages are used as weights):

$c_v = 0.595–603/T$ Cal/kg, where T is the absolute temperature (°K).

The heat of explosion is 910 Cal/kg; the explosion temperature t_e can therefore be calculated with the equation $t_e = 910/c_v$. As the unknown value t_e is expressed in °C, the equation is rewritten as:

$t_e = 910/[0.595–603/(t_e + 273)]$

The equation gives $t_e = 2441$ °C; the explosion temperature T (°K) is: $2441 + 273 = 2714$ °K.

The loading density is considered being 800 kg/m³ (0.8 kg/l), and the number of gaseous gram-molecules (per 800 kg of explosive) is calculated from the composition of the fumes:

	Weighted %	Mass (kg)	n ° gram-molecules
CO_2 (p.m. 44)	17.3	138.4	3,063.6
N_2 (p.m. 28)	33.1	264.8	7,975.8
H_2O (p.m. 18)	49.6	396.8	22,044.4
Total	100.0	800.0	33,083.8

The reduction of covolume is obtained from the diagram in Figure 2.17: for a loading density of 0.8 kg/L (1.25 L/kg), it is 0.73 L/kg, and therefore, for 800 kg, the deduction amounts to 0.73 × 800 = 584 L = 0.584 m³ (the "free" volume is then 1–0.584 = 0.416 m³).

The pressure, in Pa, can be calculated as:

$$P = 1\,794\,709\,464 \text{ Pa} = 1794.7 \text{ MPa} (\sim 18 \text{ kbar}).$$

This result, obtained in theoretical conditions, is to be considered conservative; practically, values reported in the literature vary from 12 to 15 kbar (Lee et al., 1968; Bdzil et al., 2002; Jackson et al., 2011; Jackson, 2017).

The old *Cal* unit was used in place of the International System unit (MJ), as it can be useful to split the thermal aspects from the mechanical ones.

2.8.3.2 The "specific pressure" (explosive strength) and the specific volume of the gases

The so-called "specific pressure" is rarely used today to characterize and compare explosives: it is the pressure that would be developed by 1 kg of explosive mass exploding in 1 l of volume if the fumes followed the ideal gas law. This pressure can be easily calculated from the explosion reaction (Jackson, 2017).

Although it is only roughly related to the pressure reached in the explosion, it can still be useful for the design of charges in cases where the explosive has a very low load density, which can happen in splitting works; in fact, at low load densities, the ideal gas law is still an acceptable approximation of the behavior of the fumes.

For ANFO, the "specific pressure" is about 930 MPa. At very low loading densities, for example, up to 0.1 kg/l (100 kg/m³), the pressure can be approximately calculated by multiplying the "specific pressure" by the loading density.

The "specific volume of gas" is the number of normal liters of gas (liters reduced at unitary atmospheric pressure and at room temperature, but conventionally also considering the water at a gaseous state) carried out by 1 kg of explosives through the theoretical explosion reaction. It can be calculated by multiplying by 22.4 (the number of normal liters occupied by a gram-molecule of gas under "normal" conditions) the number of gram-molecules of gas developed by 1 kg of explosives. For ANFO, it corresponds to 926 normal liters/kg.

The "specific volume" has no use for the purposes of comparing the efficiency of explosives because it does not consider the energetic aspects of the explosion nor the reaction rate.

However, it can be used to evaluate, in underground works, the volume of fumes to be disposed of through the ventilation system. For industrial explosives, the specific volume of gas is in the order of 1,000 normal liters/kg (Smith and VanNess, 1959; Britton et al., 1984; Sanchidrián and López, 2006).

2.8.3.3 Calculation of the "quasi-static" pressure of the gases starting from the "specific pressure" and the "loading density"

The "specific pressure" is simply the value of the *nRT* product, where n is the number of gram-molecules of gas developed by 1 kg of explosives.

According to the empirical graph in Figure 2.23, it can be easily stated that the relationship between the "effective" pressure of the gases and the specific pressure must be solely a function of the loading density; if ρ is the loading density (kg/L), the ratio can be simply expressed by: $\rho/(1-\rho\alpha)$ as plotted in Figure 2.18.

Table 2.1 (Berta, 1990) shows the "specific pressure" values for some industrial explosives (Italian), alongside various other characteristics.

By way of example, starting from these data and with reference to the graph in Figure 2.18, the gas pressure is calculated for some explosives (Table 2.2) at arbitrary loading densities (but in line with practice).

2.8.4   Detonation pressure

If direct measurements of the pressure of the explosion gases are difficult and unreliable, the direct measurement of the detonation pressure is practically impossible. The only available values available for detonation pressures are obtained with calculations.

Table 2.1 "Specific pressure" values of some industrial explosives (Pravisani).

Explosive (Italian commercial names)	Density (kg/m³)	Specific energy (MJ/kg)	Detonation velocity (m/s)	Specific pressure (MPa)	Detonator sensitivity (n°)	Gap sensitivity (m)	Fumes (*)	Water resistance
GOMMA A	1,550	6.74	7,500	1,246	8	>0.10	2 T	Yes
GELATINA 1	1,450	4.52	6,550	1,007	8	>0.10	T	Yes
GELATINA 2	1,420	4.44	6,100	954	8	>0.10	T	Yes
SISMIC 2	1,550	4.00	6,600	878	8	0.16	N	Yes
IDROPENT 2	1,550	7.47	7,900	1,327	8	0.25	N	Yes
PROFIL X	1,200	2.66	3,240	343	8	>0.10	T	Yes
TUTAGEX 210	1,150	3.52	4,200	828	8	>0.02	0.1 T	Yes
TUTAGEX 110	1,150	2.79	4,000	688	8	>0.02	0.1 T	Yes
VULCAN 3	1,050	3.90	4,500	899	8	0.10	1.5 T	No
CAVA EXTRA 2	1,050	4.31	4,550	966	8	0.10	N	No
CAVA 1	1,000	4.16	3,800	930	8	0.05	N	No
ANFO	800	3.66	2,300	813	10	0.02	1.5 T	No

(*) Times required for fumes clearing, at the same environmental conditions, referring to high-density straight dynamite 1 (percentage of dynamite higher than 60%), assumed to be equal to T; N: unsuitable for underground use.

Table 2.2 Gas pressure values for some industrial explosives.

Explosive (Italian commercial names)	Loading density (kg/l)	Gas pressure/specific pressure	Specific pressure (MPa)	Gas pressure (MPa)
GELATINA 2	1.4	6	954	5,724
TUTAGEX 210	1	3	828	2,487
VULCAN 3	1	3	899	2,697
ANFO	0.8	2	813	1,626
ANFO	0.7	1	813	813

Source: Berta (1990)

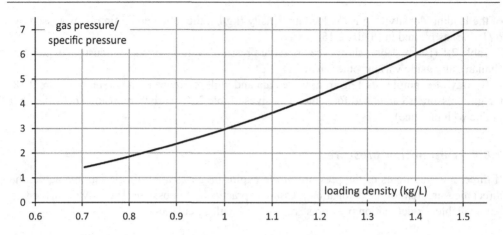

Figure 2.18 Correlation (derived from Figure 2.17) between the relationship "gas pressure/ specific pressure" and the loading density.

The detonation pressure p_d is defined by the product of the detonation velocity V_d, the density of the explosive ρ, and the particle velocity of the "fumes" W:

$$p_d = \rho \, V_d \, W$$

The scheme in Figure 2.19 provides an explanation of the concept: a cylindrical charge of section S is detonating; considering an ideal plane orthogonal to the direction of the detonation, at the time Δt, a motionless mass of explosive M, equal to $V_d \Delta t \rho S$, turns into an equal mass of fumes of speed W.

To block the velocity V_d on that plane at the instant Δt in which the fumes propagate, acquiring the velocity W, it would be necessary to apply an impulsive force F, so that $F \Delta t = M \, W$. This means: $F \Delta t = V_d \Delta t \rho SW$, that is, $F = V_d \rho SW$. Being the pressure, the ratio between the force and the surface on which it is applied can be written as $P_d = F/S = \rho \, V_d \, W$, where P_d is the detonation pressure.

2.8.4.1 Detonation pressure values

Density and detonation velocity are easily measurable; instead, the measurement of W is problematic. It has been proved by indirect and/or theoretical determinations (quite complicated) that V_d is proportional to W (for "logical" reasons, W is less than V_d, but it is difficult to quantify the difference). Admitting the proportionality between V_d and W, the expression becomes:

$$p_d = K \rho \, V_d^2$$

There is some disparity on the most suitable value to assign to K: different authors indicate values between 1/6 and 1/3 (Cook et al., 1962; Coleburn, 1964; Sheffield and Blomquist, 1984; Held, 1987; Cooper, 1996; Loboiko and Lubyatinsky, 2000; Green and Lee, 2006). The most adopted value is 1/4, and the detonation pressure is usually computed as:

$$p_d = 0.25 \, \rho \, V_d^2$$

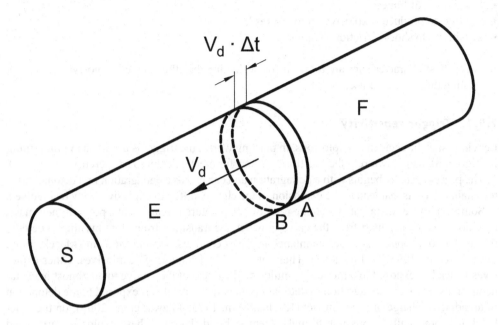

$V_d \cdot \Delta t$

F

V_d

E

B   A

S

Figure 2.19 Illustration of the concept of detonation pressure, understood as the pressure to
be applied to block the detonation in progress. A: point reached by the detonation
reaction at the instant *t*; B: point reached by the detonation reaction at the instant
t + Δ*t*; F: fumes (reaction products, under pressure); E: undetonated explosive, at
ordinary pressure; V_d: detonation velocity; S: cross section of the charge.

The formula, in SI units, gives p_d in Pa (N/m²), if ρ is expressed in kg/m³ and V_d in m/s; to obtain
the value in MPa, the result must be divided by 10^6.

For straight gelatines, p_d is of about 80 kbar, while the "quasi-static" pressure is 30–40 kbar;
for AN-TNT based explosives, P_d = 40–50 kbar, against a "quasi-static" pressure between 20
and 30 kbar.

2.9 Other important characteristics of explosives

The characteristics examined up to now (reaction velocity, specific energy, and explosion pressures) are the most important, as they specify the contribution of a given charge in terms of
pressure, duration of the mechanical action, and efficiency. They are equivalent to the "plate
data" of an engine. However, many other characteristics must be known to correctly select and
use explosives. They are equivalent to the "instruction manual" of the engine.

A list and a brief discussion are provided below:

- trigger sensitivity;
- critical diameter;
- critical density;
- gap sensitivity;
- water resistance;

- healthiness of fumes;
- aptitude to produce "explosive atmospheres";
- sensitivity to shocks, friction, heating.

Some of these characteristics are relevant for evaluating the efficiency of explosives, others for evaluating their safety of use.

2.9.1 Trigger sensitivity

Leaving aside the accidental explosions, in blasting works, the trigger is usually a thermal stimulus (ignition), and only rarely the reaction can start for mechanical impact (percussion triggers).

The process usually begins with a deflagration and can end as a deflagration or a detonation; in the chain of events that lead to the explosion of the charge, different explosives may be involved.

Some explosives deflagrate when the ignition temperature is reached at a point of their mass; they do not detonate, apart from the cases in which the transition from slow-burning to detonation is due to a shock that arises spontaneously in a confined burning medium (Maček, 1959; Luebcke et al., 1995; Frolov, 2008). These are called "deflagrating" explosives. Other explosives instead, if exposed to a thermal stimulus, deflagrate, but they move to an almost instantaneous transition from deflagration to detonation; they detonate also if exposed to the detonation of an adjacent charge or to an impact (Kirshenbaum, 1975; Talawar et al., 2006); on the other hand, it is practically impossible to make them stably deflagrate. These explosives are called "primary", and in practice, they are used only in detonators. Most of the explosives used in rock blasting belong to the category of "secondary" detonating explosives. These kinds of explosives deflagrate when locally exposed to the ignition temperature, but generally, the reaction does not evolve to detonation; this transition, however, can occur in some cases. It is used, for example, to realize the so-called detonators without primary explosive (Ziegler, 1987; Shen and Ma, 2009; Jian-guo et al., 2012; Matyáš and Pachman, 2013; Persson et al., 2018). The detonation is instead obtained by exposing these explosives to the action of the detonation of another charge.

The secondary detonating explosives are still divided into two categories: "sensitive to the detonator" and "low sensitive". The first group pertains to explosives whose regular detonation is initiated by a small primary charge (about 2 g) contained in a detonator. The second group, low-sensitive explosives (blasting agents), require the action of a high-sensitive auxiliary charge, called a "booster", to detonate regularly.

The classification is visually illustrated in Figure 2.20.

Quantitatively, the trigger sensitivity of an explosive is defined by the "number", in the detonator sensitivity scale (or "power") capable of initiating it safely.

2.9.2 Critical diameter

Explosives are generally available in the shape of cylindrical charges: they are triggered at one end, and the detonation should propagate up to the other end, but it can only propagate if enough energy is transferred from the detonation front to the explosive which has not yet reacted, to activate the reaction of a further layer of explosive. A given amount of energy is inevitably transmitted outside the boundary of the reaction zone, which in return does not partake in the extension of the process (Figure 2.21). The ratio of the circumference to the cross section increases as the section decreases; for this simple reason, since the loss is proportional to the circumference, its incidence (%) increases with the decrease of the diameter of the charge: for any explosive, there

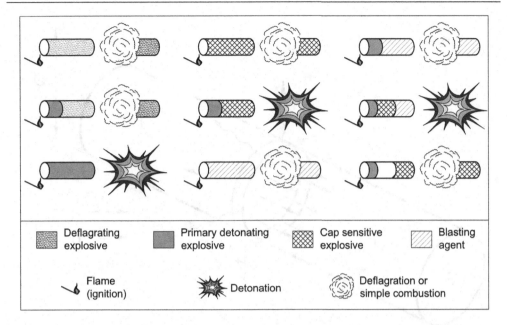

Figure 2.20 Modes of initiation and types of reactions that can be obtained for different combinations of charges placed in succession. It is important to note that: (a) for certain secondary detonating explosives, the deflagration is no faster than ordinary combustion; (b) the primary charges considered are those contained in ordinary detonators (a few grams), but large primary charges could also initiate the low-sensitive explosives. The cap-sensitive explosive is a secondary "sensitive" detonating explosive, while the blasting agent is a secondary low-sensitive explosive.

is, therefore, a "critical diameter" (some millimeters for some explosives, some centimeters for others), below which the detonation, triggered at one end of the charge, cannot propagate, as the energy intended to keep the process has become insufficient (Dremin and Trofimov, 1965; Engelke, 1983; Petel et al., 2007).

The transversal dispersion is also an increasing function of the thickness of the reaction zone, and this explains the fact that the "ideally" detonating explosives (with a very small reaction zone thickness) have a smaller critical diameter than the "non-ideal" explosives (Jaffe and Price, 1962; Price, 1967; Kobylkin, 2009). For similar reasons, the diameter also affects the detonation velocity: charges of the same explosive having a bigger diameter detonate at a higher speed.

2.9.3 Charge density

An explosive can become unsuitable for the propagation of detonation (and insensitive to trigger) due to overly high compaction: the phenomenon is due to the so-called "exceeding the critical density". In fact, there is a special density value (proper for each type of explosive) that has not to be exceeded for correct employment. With reference to the same phenomenon, the term "deactivation pressure" is also used, as the density of the explosive increases with the pressure (Price, 1967; Sil'vestrov, 2006).

Intuitively, this effect can be explained by a simple consideration: to guarantee the propagation of the detonation, it is necessary that the explosive has a certain deformability, allowing it

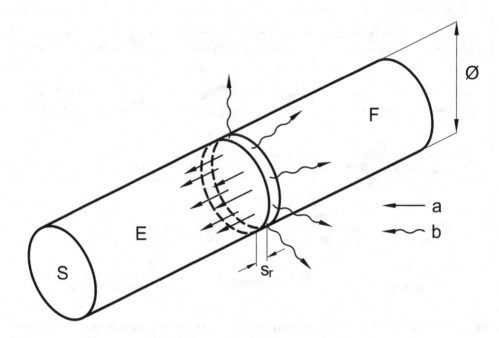

Figure 2.21 Effect of the diameter of the charge on the lateral (or peripheral) dispersion of energy. Being $S = \dfrac{\pi d^2}{4}$ and the circumference $C = \pi d$, the C/S ratio obviously increases as the diameter decreases, as well as the ratio of the energy wasted, proportional to C, to that transmitted into the yet-to-be-reacted charge, proportional to S.

E: explosive; F: fumes; s_r: thickness of the reaction zone; S: cross section of the charge; D: diameter of the charge; a: energy transmitted to the still unreacted charge; b: energy wasted transversely to the charge axis.

to extract energy from the advancing shock wave. An example can be given by considering the different heat that can be obtained by hammering on a piece of lead, which is very deformable, or on a piece of steel.

The danger of incomplete detonation due to excessive compaction of the charge is to be considered especially when blasting agents are used, where the density limit can be easily exceeded when the loading is poorly controlled. For example, the critical density of ANFO is about 1.2 kg/L. With the use of explosives of this type, increasing the compaction with the aim of introducing more charge into the hole to obtain a greater effect involves the risk of failure (Nie et al., 1993; Fleetwood et al., 2012; Fabin and Jarosz, 2021). It has been mentioned that the ANFO has an explosion pressure higher than 10 kbar; at this pressure, it would certainly reach a density greater than 1.2 kg/L (the density of ammonium nitrate in crystals is 1.725). The charge is not deactivated during the explosion, because the detonation wave proceeds at a supersonic speed, that is, higher than the speed at which a pressure variation could proceed (the pressure variation corresponds to a density variation). In underwater excavation works at great depths or, in any case, under high water levels (e.g., in explosions for geoseismic research), the external pressure may be sufficient to compact many explosives beyond the critical density. In this case, it may be necessary to use the explosives in airtight envelopes, which provide protection against the compacting effect of external pressure.

2.9.4 *Gap sensitivity*

The gap sensitivity is the capacity to transmit detonation from one initiator cartridge (the donor) to another (the receptor) located at a certain distance away; this effect is also designated as "flashover"(Berta, 1990). Obviously, the distance depends on the size of the charges; for the same size, it depends on the type of explosive, and again for the same size and type of explosive, on the way the test is performed (the most used procedures are shown in Figure 2.22). The test

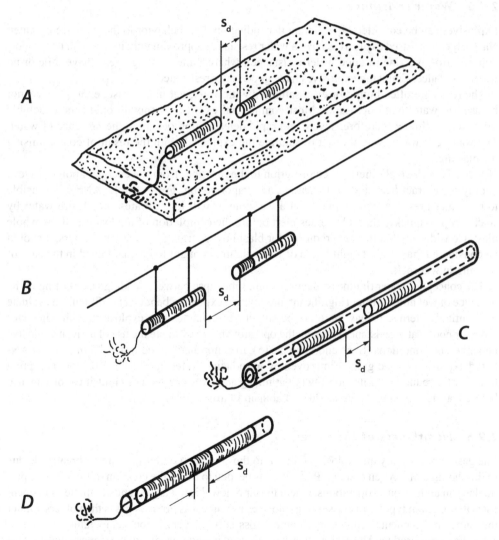

Figure 2.22 Different methods of determining the gap sensitivity d_c. A: charges arranged on a sand bed; B: free suspended charges; C: charges confined in a tube, with a cavity full of air; D: charges spaced by inert disks, for example, cardboard.

is quite challenging, as the identification of the gap sensitivity requires numerous attempts and statistical validation.

According to an alternative definition of the gap sensitivity, d_c is the distance that the shock wave generated by the explosion in an inert medium (no longer, therefore, a detonation front) can cover, remaining suitable for triggering the detonation of another charge, that is to "re-transform itself" into detonation front. A high gap sensitivity can be wanted (it ensures the complete detonation of a series of cartridges placed in a blast hole even if they are not in mutual contact) or unwanted (it can induce the simultaneous detonation of two charges that instead should detonate at different times).

2.9.5 Water resistance

Explosives can be considerably different depending on their behavior in the presence of water. Nitro-glycerine-based explosives have a water resistance approximately in line with the proportion of nitro-glycerine (a good water resistance), whereas the blasting agents have little or no water resistance. Emulsion explosives have good water resistance.

The rocks are often humid or even saturated with water, yet in any case, explosives cannot be used in watertight enclosures. Most commercial explosives contain oxidizers as soluble salt, usually nitrate; therefore, severe and rapid deterioration occurs in the presence of water. The water can wash away the salt or saturate the explosive, with the result of compromising the blasting.

Water is explosively inert: its contribution to the reaction, from the energetic point of view, is only to subtract heat due to its heating and vaporization. For example, ANFO is sensible to the water presence: if it is placed in a wet hole, it will lose nitrates and absorb water by soaking up so quickly that it becomes inert before the completion of the loading of the whole blast. Considering that the time required for blast hole charging is a matter of hours, it should be considered that at least eight hours of water tightness ought to be guaranteed in the case of blasting in wet rock.

The concept of water tightness (the maximum time lapse between loading and blasting in the presence of wet holes without significant alteration in explosive behavior) is difficult to evaluate in quantitative terms. It may therefore be sufficient to characterize explosives with adjectives (good or poor water resistance) making the operator able to evaluate the use of a given explosive under certain conditions (Shao and Feng, 2015). In contradiction, the water tightness of an AN-based explosive can be greatly improved by introducing water into its composition, accepting however the reduction of its mobility (gelatinizing additives or adequate emulsification can also be used, as described in Ali et al., 2021; Fabin and Jarosz, 2021).

2.9.6 Healthiness of the fumes

The gases produced by an explosion (similar to those of any combustion) are unbreathable due to the shortage of oxygen content. Precaution is needed to avoid the creation of toxic gases, particularly in underground operations, where the air renewal is not as fast and complete as in opencast sites. Certain types and classes of explosives produce less noxious gases than others. Under the same environmental conditions (tunnel cross section, ventilation, temperature, humidity, etc.), the fumes produced by different explosives used in underground mining operations present different atmospheres in the fume. Therefore, the dispersal times and the obligatory wait before returning to the site of operation vary according to the explosive used (Berta, 1990).

The formation of carbon monoxide should always be avoided. Its formation is due to incomplete oxidation, and for this reason, all explosives to be used underground must contain an excess of oxygen (the null balance does not provide a sufficient guarantee, as the actual explosion reaction may deviate from the theoretical one and because some tolerance must be allowed in the formulation). Other feared components are nitrogen oxides: their production is generally favored by a surplus of oxygen.

Adequate ventilation is, in any case, the most important single factor in ensuring a proper atmosphere after blasting.

Of course, a simple desk check of the theoretical explosion reaction is not enough, but analysis of the fumes produced by test explosions must also be performed.

2.9.7   Behavior in explosive atmospheres

Mixtures of air with flammable gases or flammable powders, within certain composition limits, maybe "explosive", meaning that a flare, spark, or another suitable thermal stimulus may cause the mixture explosion or detonation. These mixtures are certainly not "bright" explosives: density is very low, just over 1 kg/m^3, and the explosion pressure can reach a few bars, at most a dozen, against the thousands of bars of solid explosives; anyway, the volume they occupy is very big, and therefore, the destructive effect of the explosion can be significant; moreover, in case of an underground site (tunnel, mine), the consequences of an explosion could be brutal for operators.

The most common types of explosive atmospheres are the air–methane (grisou) and air–coal dust mixture, which represent a typical danger in coal mines but can also originate in other sites; other flammable gases, vapors, or dust can also produce explosive atmospheres. If explosives are used in an explosive atmosphere, the detonation of the charges will potentially induce that of the surrounding atmosphere, with disastrous consequences (Lupu et al., 2014). In these cases, the use of explosives should be prohibited, or "safety" explosives should be used (Wang et al., 2013): they are explosives at low explosion temperatures and high reaction rates which, when used properly, do not trigger the explosion reaction of the atmosphere. The explosion temperature must not exceed 1400 °C (indeed, this temperature is sufficient to ignite the air/methane mixture, which is the reference explosive atmosphere, but to achieve this effect, it should remain for a much longer time than the duration of the explosion); it is also required that the formulation contains an adequate excess of oxygen, to prevent the fumes from containing incomplete oxidation products, which are actually combustible and can ignite after the charge explodes (Cooper and Kurowski, 1997). To qualify an explosive as "safe", it is not enough to study the explosive reaction, but it is necessary to carry out severe tests of use in explosive atmospheres artificially produced in experimental tunnels.

The most common safety explosives are like dynamites (the presence of nitro-glycerine is necessary to ensure rapid detonation) but added with inert substances that act as "refrigerants" (usually chlorides). They have specific energy and explosion pressures much lower than normal explosives.

2.9.8   Sensitivity to shocks, friction, and heating

The indicators of these characteristics, which are important for guaranteeing the safety of transport, handling, and storage of explosives, are given by the results of the "shearwater" test (test of sensitivity to shocks, carried out by determining the height of fall of a mass of steel on a small charge forced between two surfaces of pure steel, which causes the explosion of the charge itself), of the friction test of a small charge between two rough surfaces (steel or ceramic material) according to standardized methods, from the determined ignition temperature by heating a small charge on a plate which can be brought to a known temperature with good accuracy or with similar tests (Andersen, 1981; Bennett et al., 1998; Peterson et al., 2007). All these tests represent a much more severe treatment for the explosive than it would have in practice (before, of course, being deliberately detonated).

References

Afanasenkov, A.N., 2004. Strength of explosives: Trauzl test. Combustion, Explosion and Shock Waves, 40(1), pp. 119–125.

Agrawal, H. and Mishra, A.K., 2018a. A study on influence of density and viscosity of emulsion explosive on its detonation velocity. Model Meas Control C, 78(3), pp. 316–336.

Agrawal, H. and Mishra, A.K., 2018b. A study on influence of density and viscosity of emulsion explosive on its detonation velocity. AMSE Journals-AMSE IIETA Publication, 78(3), pp. 316–336.

Ali, F., Pingua, B.M., Dey, A., Roy, M.P. and Singh, P.K., 2021. Surface functionalized ammonium nitrate prills with enhanced water resistance property: Characterizations and its application as commercial explosives. Propellants, Explosives, Pyrotechnics, 46(1), pp. 78–83.

Andersen, W.H., 1981. Role of the friction coefficient in the frictional heating ignition of explosives. Propellants, Explosives, Pyrotechnics, 16(1), pp. 17–23.

Bauch, E. and Lempp, C., 2004. Rock splitting in the surrounds of underground openings: An experimental approach using triaxial extension tests. Engineering Geology for Infrastructure Planning in Europe. Lecture Notes in Earth Sciences, vol. 104. Springer, Berlin, Heidelberg.

Bdzil, J.B., Aslam, T.D., Catanach, R.A., Hill, L.G. and Short, M., 2002. DSD front models: Nonideal explosive detonation in ANFO. Proceedings of the 12th International Detonation Symposium, San Diego, CA, USA, pp. 409–417.

Bennett, J.G., Haberman, K.S., Johnson, J.N. and Asay, B.W., 1998. A constitutive model for the non-shock ignition and mechanical response of high explosives. Journal of the Mechanics and Physics of Solids, 46(12), pp. 2303–2322.

Berger, J. and Viard, J., 1962. Physique des explosifs solides. Dunod Ed., Paris, France.

Berta, G., 1990. Explosives: An Engineering Tool. Italesplosivi, Milano.

Berthoud, G., 2000. Vapor explosions. Annual Review of Fluid Mechanics, 32, pp. 573–611.

Beyling, C. and Drekopf, K., 1936. Sprengstoffe und Zündmittel. Springer Ed, Berlin, Germany, 465 pp.

Bragg, W.L. and Williams, E.J., 1934. The effect of thermal agitation on atomic arrangement in alloys. Proceedings of the Royal Society of London. Series A, Containing Papers of a Mathematical and Physical Character, 145(855), pp. 699–730.

Britton, R.R., Skidmore, D.R. and Otuonye, F.O., 1984. Simplified calculation of explosives-generated temperature and pressure. Mining Science and Technology, 1(4), pp. 299–303.

Chalon, F., 1911. Les explosifs modernes. Béranger Ed., Paris, France.

Chiappetta, R.F., 1993. Continuous velocity of detonation measurements in full scale blast environments. Proceedings of the International Seminar on Rock Blasting (Abattage des Roches à l'Explosif), Alès, France, pp. 27–74.

Chiappetta, R.F., 1998. Blast monitoring instrumentation and analysis techniques, with an emphasis on field applications. Fragblast-International Journal for Blasting and Fragmentation, 2(1), pp. 79–122.

Coleburn, N.L., 1964. Chapman – Jouguet pressures of several pure and mixed explosives. NOLTR 64–58, United States Naval Ordnance Laboratory, Maryland, USA. DTIC Accession Number AD0603540.

Cook, M.A., 1958. The Science of High Explosives. Reinhold Publ. Co., New York, USA.

Cook, M.A., Keyes, R.T. and Ursenbach, W.O., 1962. Measurements of detonation pressure. Journal of Applied Physics, 33(12), pp. 3413–3421.

Cooper, P.W., 1996. Explosives Engineering. Wiley-WCH Inc, New York.

Cooper, P.W. and Kurowski, S.R., 1997. Introduction to the Technology of Explosives. John Wiley & Sons, Albuquerque, NM, USA, 209 pp.

Cowan, R.D. and Fickett, W., 1956. Calculation of the detonation properties of solid explosives with the Kistiakowsky-Wilson equation of state. The Journal of Chemical Physics, 24(5).

Davis, T.L., 2016. The Chemistry of Powder and Explosives. Originally published 1943, J. Wiley, Hoboken, NJ, USA, 490 pp.

Deal, W.E., 1957. Measurement of Chapman-Jouguet pressure for explosives. The Journal of Chemical Physics, 27(3).

Dick, R.A., Fletcher, L.R. and D'Andrea, D.V., 1986. Explosives and Blasting Porcedures. ABA Publishing Company – "The Blaster's Library", Washington, DC, USA, 105 pp.

Dremin, A.N. and Trofimov, V.S., 1965. On the nature of the critical diameter. Proceedings of International Symposium on Combustion, 10(1), pp. 839–843. Elsevier Ed.

Elshenawy, T., Seoud, A.A. and Abdo, G.M., 2019. Ballistic protection of military shelters from mortar fragmentation and blast effects using a multi-layer structure. Defence Science Journal, 69(6), pp. 538–544.

Engelke, R., 1983. Effect of the number density of heterogeneities on the critical diameter of condensed explosives. The Physics of Fluids, 26(9), pp. 2420–2424.

Euler, L., 1760. Lettere ad una principessa tedesca. Lett. 13. Boringhieri Ed., Torino, Italy.

Fabin, M. and Jarosz, T., 2021. Improving ANFO: Effect of additives and ammonium nitrate morphology on detonation parameters. Materials, 14(19), p. 5745.

Feldgun, V.R., Karinski, Y.S., Ebri, I. and Yankelevsky, D.Z., 2015. Prediction of the quasi-static pressure in confined and partially confined explosions and its application to blast response simulation of flexible structures. International Journal of Impact Engineering, 90(2016), pp. 46–60.

Fleetwood, K.G., Villaescusa, E. and Eloranta, J., 2012. Comparison of the non-ideal shock energies of sensitised and unsensitised bulk ANFO-emulsion blends in intermediate blasthole diameters. Proceedings of the Thirty-Eighth Conference on Explosives and Blasting Technique, Nashville, TN, USA.

Frolov, S.M., 2008. Fast deflagration-to-detonation transition. Russian Journal of Physical Chemistry, 2, pp. 442–455.

Fujishiro, T., Iwata, K., Kikuchi, T., Yoshihara, F. and Hoshi, T., 1974. Japan Atomic Energy Research Inst., Tokyo. OSTI Identifier: 4212369, Report Number; JAERI-M-5861. Technical report.

Gogulya, M.F., Makhov, M.N., Dolgoborodov, A.Y. et al., 2004. Mechanical sensitivity and detonation parameters of aluminized explosives. Combustion, Explosion, and Shock Waves 40, pp. 445–457.

Gordon, W.E., Reed, F.E. and Lepper, B.A., 1955. Lead-block test for explosives. Industrial and Engineering Chemistry, 47(9), pp. 1794–1800.

Green, L.G. and Lee, E.L., 2006. Detonation pressure measurements on PETN. Proceedings of the 13th International Detonation Symposium, Norfolk, VA, USA, July 23–28.

Green, S.P., Wheelhouse, K.M., Payne, A.D., Hallett, J.P., Miller, P.W. and Bull, J.A., 2020. Thermal stability and explosive hazard assessment of diazo compounds and diazo transfer reagents. Organic Process Research & Development, 24, pp. 67–84.

Held, M., 1987. Determination of the chapman – Jouguet pressure of a high explosive from one single test. Defence Science Journal, 37(1), pp. 1–9.

Jackson, S.I., 2017. The dependence of Ammonium-Nitrate Fuel-Oil (ANFO) detonation on confinement. Proceedings of the Combustion Institute, 36(2), pp. 2791–2798.

Jackson, S.I., Kiyanda, C.B. and Short, M., 2011. Experimental observations of detonation in Ammonium-Nitrate-Fuel-Oil (ANFO) surrounded by a high-sound-speed, shockless, aluminum confiner. Proceedings of the Combustion Institute, 33(2), pp. 2219–2226.

Jaffe, I. and Price, D., 1962. Determination of the critical diameter of explosive materials. ARS Journal, 32(7), pp. 1060–1065.

Jian-guo, D., Hong-hao, M. and Zhao-Wu, S., 2012. A technique for highly precise and safe delay detonator without primary explosive. Electrical Measuring Instruments and Measurements, p. 41.

Kirshenbaum, M.S., 1975. Functional circuit parameter approach to the electrostatic sensitivity of primary explosives. International Conference on "Research on Primary Explosives" Editors, ERDE, Waltham Abbey, England.

Ko, Y.H., Kim, S.J., Baluch, K. and Yang, H.S., 2017. Study on blast effects of stemming materials by Trauzl lead block test and numerical analysis. Explosives and Blasting, 35(4), pp. 19–26.

Kobylkin, I.F., 2009. Critical detonation diameter of highly desensitized low-sensitivity explosive formulations. Combustion, Explosion, and Shock Waves, 45(6), pp. 732–737.

Langefors, U. and Kihlström, B., 1967. The Modern Technique of Rock Blasting, 2nd Ed. Almqvist & Wiksell Ed., Stockolm, Sweden.

Lee, E.L., Horning, H.C. and Kury, J.W., 1968. Adiabatic Expansion of High Explosive Detonation Products. Technical Report UCRL-50422, Lawrence Livermore National Laboratory, Livermore.

Leiper, G.A., 1989. The behaviour of non-ideal explosives in the Ballistic mortar. Journal of Energetic Materials, 7(4–5), pp. 381–404.

Loboiko, B.G. and Lubyatinsky, S.N., 2000. Reaction zones of detonating solid explosives. Combustion, Explosion, and Shock Waves (Engl. Transl.), 36(6), pp. 716–733.

Luebcke, P.E., Dickson, P.M. and Field, J.E., 1995. An experimental study of the deflagration-to-detonation transition in granular secondary explosives. Proceedings of the Royal Society – Mathematical, Physical and Engineering Sciences, pp. 439–448.

Lupu, L., Ghicioi, E., Jurca, A. and Paun, F., 2014. Ensuring security and environmental safety at blasting workplaces. Environmental Engineering and Management Journal, 13(6), pp. 1517–1522.

Maček, A., 1959. Transition from deflagration to detonation in cast explosives. The Journal of Chemical Physics, 31, p. 162.

Macek, A., 1962. Sensitivity of explosives. Chemical Reviews, 62(1), pp. 41–63. ACS Publications.

Mainiero, R.J., 1997. A technique for measuring toxic gases produced by blasting agents. Proceedings of the Twenty Third Annual Conference on Explosives and Blasting Technique, Las Vegas, NV, February 2–5.

Matyáš, R and Pachman, J., 2013. Primary Explosives. Springer Ed., Heidelberg, pp. 325–328.

Mertuszka, P., Cenian, B., Kramarczyk, B. and Pytel, W., 2018. Influence of explosive charge diameter on the detonation velocity based on emulinit 7L and 8L bulk emulsion explosives. Central European Journal of Energetic Materials, 15(2), pp. 351–363.

Mertuszka, P. and Pytlik, M., 2019. Analysis and comparison of the continuous detonation velocity measurement method with the standard method. High Energy Materials, 11(2), pp. 63–72.

Mishra, A.K., Agrawal, H. and Raut, M., 2019. Effect of aluminum content on detonation velocity and density of emulsion explosives. Journal of Molecular Modeling, 25, p. 70.

Nie, S., Deng, J. and Persson, A., 1993. The dead-pressing phenomenon in an ANFO explosive. Propellants, Explosives, Pyrotechnics, 18(2), pp. 73–76.

Olmsted, R.M., Chiappetta, R.F. and Palangio, T., 1998. New HRS-1 borehole inspection video camera system. Proceedings of the BAI's Seventh High-Tech Seminar on State-of-the-Art Blasting Technology Instrumentation and Explosives Applications, Orlando, FL, USA, pp. 735–753.

Östmark, H., Wallin, S. and Ang, H.G., 2012. Vapor pressure of explosives: A critical review. Propellants, Explosives, Pyrotechnics, 37, pp. 12–23.

Palangio, T.C., Maerz, N.H. and Franklin, J.A., 1997. WipFrag and WipJoint to measure, record and predict blast results. Proceedings of the BAI's Seventh High-Tech Seminar on State-of-the-Art Blasting Technology, Instrumentation and Explosives Applications, Orlando, FL, USA, pp. 487–510.

Peralta, D., Paterson, N.P., Dugwell, D.R. and Kandiyoti, R., 2001. Coal blend performance during pulverised-fuel combustion: Estimation of relative reactivities by a bomb-calorimeter test. Fuel, 80(11), pp. 1623–1634.

Persson, P.A., Holmberg, R. and Lee, J., 1994. Rock Blasting and Explosives Engineering. CRC Press, Boca Raton, FL, USA.

Persson, P.A., Holmberg, R. and Lee, J., 2018. Rock Blasting and Explosives Engineering. CRC Press, Boca Raton, FL, USA, 540 pp.

Petel, O.E., Mack, D., Higgins, A.J., Turcotte, R. and Chan, S.K., 2007. Minimum propagation diameter and thickness of high explosives. Journal of Loss Prevention in the Process Industries, 20(4–6), pp. 578–583.

Peterson, P.D., Avilucea, G.R., Bishop, R.L. and Sanchez, J.A., 2007. Individual contributions of friction and impact on non-shock initiation of high explosives. In AIP Conference Proceedings, 955(1), December, pp. 983–986. American Institute of Physics.

Price, D., 1967. Contrasting patterns in the behavior of high explosives. Proceedings of the International Symposium on Combustion, 11(1), pp. 693–702. Elsevier Ed.

Reddi, K.K., 2018. Measurement of Velocity of Detonation (VOD) of Explosives Using Dautriche and Electronic Methods. MTech thesis. Deposited by IR Staff BPCL – Ethesis@NIT Rourkela, India.

Rowland, J.H. and Mainiero, R.J., 2000. Factors affecting ANFO fumes production. Proc 26th Conf Explos Blasting Tech, Anaheim, CA, February 13–16. International Society of Explosives Engineers, Cleveland, OH, pp. 163–174.

Sanchidrián, J.A. and López, L.M., 2006. Calculation of the energy of explosives with a partial reaction model: Comparison with cylinder test data. Propellants, Explosives, Pyrotechnics, 31(1), pp. 25–32.

Satyavratan, P.V. and Vedam, R., 1980. Some aspects of underwater testing method. Propellants Explosives Journal, 5(2/3).

Shao, A.L. and Feng, S.R., 2015. Research on high-efficiency composite oil phase material for emulsion heavy ANFO. Machinery, Materials Science and Energy Engineering (ICMM 2015). Proceedings of the 3rd International Conference, Guangde Zhang Ed., World Scientific Pub., Wuhan, China, pp. 561–569.

Sheffield, S.A. and Blomquist, D.D., 1984. Subnanosecond measurements of detonation fronts in solid high explosives. The Journal of Chemical Physics, 80(8), pp. 3831–3844.

Shekhar, H., 2012. Studies on empirical approaches for estimation of detonation velocity of high explosives. Central European Journal of Energetic Materials, 9(1), pp. 39–48.

Shen, Z.W. and Ma, H.H., 2009. The key technique of highly precise and safe delay detonator without primary explosive. Proceedings of the 9th Int. Symp. on Rock Fragmentation by Blasting-Fragblast 9, Granada, Spain, J.A. Sanchidrian Ed., CRC Press, Boca Raton, FL, USA, p. 191.

Sil'vestrov, V.V., 2006. Density dependence of detonation velocity for some explosives. Combustion, Explosion and Shock Waves, 42(4), pp. 472–479.

Smith, J.M. and VanNess, H.D., 1959. Introduction to Chemical Engineering Thermodynamics, 2nd Ed. cGraw-Hill Book Co., Inc., Columbus.

Suceska, M., 2012. Test Methods for Explosives. Springer Science & Business Media, New York, NY, USA, 225 pp.

Talawar, M.B., Agrawal, A.P., Anniyappan, M., Wani, D.S., Bansode, M.K. and Gore, G.M., 2006. Primary explosives: Electrostatic discharge initiation, additive effect and its relation to thermal and explosive characteristics. Journal of Hazardous Materials, 137(2), pp. 1074–1078.

Taylor, W. and Morris, G., 1932. The absolute measurement of the available energy of high explosives by the ballistic mortar. Transactions of the Faraday Society, 28, pp. 545–557.

Tete, A.D., Deshmukh, A.Y. and Yerpude, R.R., 2013. Velocity of Detonation (VOD) measurement techniques practical approach. International Journal of Engineering and Technology, 2(3), pp. 259–265.

van Kessel, L.B.M., Arensen, A.R.J. and Brem, G., 2004. On-line determination of the calorific value of solid fuels. Fuel, 83(1), pp. 59–71.

Vermorel, O., Quillatre, P. and Poinsot, T., 2017. LES of explosions in venting chamber: A test case for premixed turbulent combustion models. Combustion and Flame, 183, pp. 207–223.

Wang, D., Zhang, D., Li, Y., et al., 2012. Experiment investigation on quasi-static pressure in explosion containment vessels. Acta Armamentarii, 33(12), pp. 1493–1497.

Wang, H.L., Xu, W.Y. and Shao, J.F., 2014. Experimental researches on hydro-mechanical properties of altered rock under confining pressures. Rock Mechanics and Rock Engineering, 47, pp. 485–493.

Wang, Y., Jiang, W., Song, D., Liu, J., Guo, X., Liu, H. and Li, F., 2013. A feature on ensuring safety of superfine explosives. Journal of Thermal Analysis and Calorimetry, 111(1), pp. 85–92.

Yoshida, T., Muranaga, K., Matsunaga, T. and Tamura, M., 1985. Evaluation of explosive properties of organic peroxides with a modified Mk III ballistic mortar. Journal of Hazardous Materials, 12(1), pp. 27–41.

Zhang, Y., Su, J., Li, Z., Jiang, H., Zhong, K. and Wang, S., 2018. Quasi-static pressure characteristic of TNT's internal explosion. Explosion and Shock Wave, 38(6), pp. 1429–1434.

Zhong, W. and Tian, Z., 2013. Numerical calculation of quasi-static pressures of confined explosions considering chemical reactions kinetic of detonation products. Explosion and Shock Waves, 33(1), pp. 78–83.

Ziegler, K., 1987. New developments in the field of firing techniques. Propellants, Explosives, Pyrotechnics, 12(4), pp. 115–120.

Chapter 3

"Operational" definitions of explosives

3.1 Introduction

As in all excavation works, the goal is to remove a given volume of rock from its natural position, yet two extreme cases can be identified:

1) The rock volume must be reduced into a "granular" material, as shown in the example of Figure 3.1, that is, a comminution effect is sought. In this case, the explosive can be defined (Definition II) as "a way to put a comminution work into the rock" (Rustan and Vutukuri, 1983; Fourney, 1993; Bozic, 1998; Kulatilake et al., 2012; Ouchterlony, 2005; Afum and Temeng, 2015; Cunningham, 1983, 1987, 2000, 2005; Shad et al., 2018; Chung and Katsabanis, 2010; Kanchibotla et al., 1999; Katsabanis and Omidi, 2015).
2) The goal is to induce fractures to isolate a given volume of sound rock, with no fragmentation, to obtain a splitting effect, as in the example of Figure 3.2.

In this case, the explosive can be defined (Definition III) as "a way to pressurize the rock locally".

Observations and explanations

Indeed, there are also intermediate situations: the two extreme examples illustrated are intended to clarify that there are at least two ways of using the explosive, which correspond to two ways of planning the work.

The "specific consumption of explosive" or "powder factor" (the ratio of the amount of charge, in kg or other units, to the blasted volume, in m^3) in the first case can be used as a parameter to calculate the charge (by knowing that × m^3 of rock have to be blasted, x is multiplied by a suitable "specific consumption" derived, for example, from the statistical analysis of a certain number from similar cases, or taken from catalogues, and the necessary charge has concurred).

In the second case, the powder factor is calculated as *a posteriori*, being a useful indicator of the unit cost of the operation. However, these points will be further developed when dealing with the design criteria of blasting works.

In addition to comminution, or to splitting, or to their combination, the blast has another effect: the displacement of the rock volume from its original position.

The displacement is usually considered a "by-product" of the main objective; however, there are also cases (so-called "casting", "throw blasting" operations, and so on), where the explosive is used for its propulsive effect: in these cases, it could be operationally defined as a transportation medium (a substitute of earth moving machines).

DOI: 10.1201/9781003241973-3

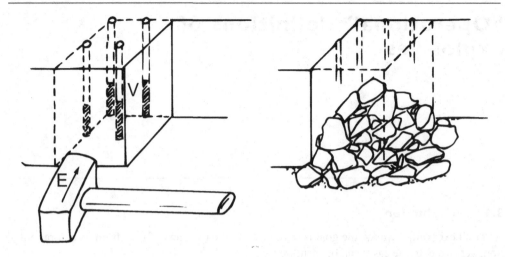

Figure 3.1 Example of blast intended to obtain rock fragmentation. The effect depends on the amount of energy *E* coerced, in relation to the volume *V* to be fragmented.

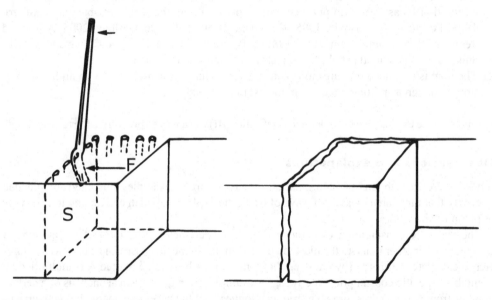

Figure 3.2 Example of a blast intended to obtain the splitting of a regular rock block. The effect depends on the extent of the applied force *F* related to the release surface *S*.

Finally, it should be recalled that, while the result is defined in a rather vague way in case 1 (it is necessary to obtain a given volume of crushed rock, but of what size?), the goal is more precisely and simply defined in case 2: for that reason, in the two cases, there are different needs for "accuracy", both in design and execution (Esena et al., 2003; Lu et al., 2016).

In fact, even though all blasting works are "controlled", the term "controlled blasting" is usually applied to cases like 2 (Jung et al., 2011; Dindarloo et al., 2015; Calder and Tuomi, 1980; Chiappetta, 2001).

3.2   Mechanical strength of rocks

When planning a blast, on a decimetric-metric scale, it would be important to have information on the mechanical behavior of rocks under impulsive and very high loads, especially under tensile and shear stresses. Moreover, it would be interesting to know both the stresses necessary to break and the work consumed before reaching the break.

On the other hand, almost all geomechanical problems are linked to stability, rather than disruption. Therefore, the conventional characterizations of rocks refer to their behavior under static loads; the most commonly used indicator is the uniaxial compressive strength, determined in the laboratory on decimetric specimens. Furthermore, the portion of the stress/strain diagram that straddles the breaking point and possibly the most interesting range for evaluating the work spent before attaining failure is the most difficult to detect and, in return, the least precisely known. For these reasons, it is impossible to derive a good indicator of blasting work from the ordinary geomechanical data that characterize the rocks (Persson and Holmberg, 1983).

In any case, Figure 3.3 gives an overview of typical values, in terms of ranges of uniaxial compressive strength (σ_R) and elastic compression modulus (E) from static tests, of the most common

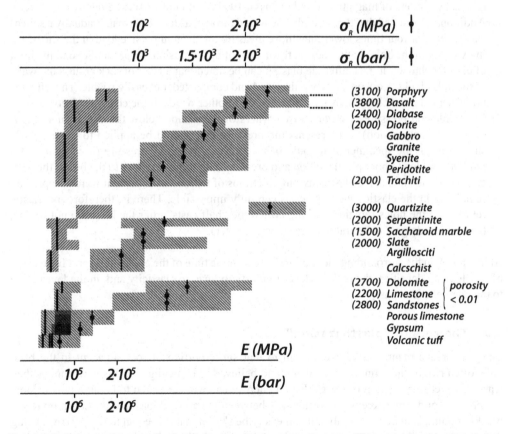

Figure 3.3 Ranges of σ_R and E of different rocks, from static tests. The considerable dispersion is not only due to rocks with different mineralogical compositions and structural characteristics that are included under the same name but also due to the different states of chemical-mineralogical alteration and mechanical integrity.

rock types (igneous, metamorphic, and sedimentary). The values are provided in bars, as the explosion pressures are traditionally expressed in this unit; the maximum value for each category is between brackets. Tensile strengths are 10–20 times lower than compressive strengths (referring to static tests again); also, the tensile elastic modulus is to some extent lower than the compressive one.

Comments and explanations

The purpose of the sketch in Figure 3.3 is, above all, "dissuasive", to induce caution in deducing the strength of a rock based on its name. Another recommendation must be made with regard to "automatic" strength assessments: in fact, the strengths measured in the laboratory on decimetric specimens are only slightly correlated to the strength of the rock mass on a metric scale. Moreover, at all scales, the anisotropy makes the "brutal" use of strength uncertain, even in static problems (not to mention dynamic ones). However, some general considerations can be made based on the data reported in Figure 3.3:

1) The explosion pressures of the most common explosives, used at loading densities of around 1 kg/l, being in the order of tens of kbar, exceed the (static) compressive strengths of rocks by one or two orders of magnitude (Szuladzinski, 1993; Edri et al., 2011; Feldgun et al., 2016). Additionally, granted that the strengths measured through a static test (load gradually applied) can be three or four times lower than those measured with an impulsive load in the surrounding vicinity of the holes in the rock, at the instant of the explosion, it goes far beyond the limit of elastic behavior. In fact, after the blast, it can be noted that a layer of rock around the walls of the holes looks "cooked": finely pulverized and compacted, but easily friable. This effect is not observed if the charge density is reduced or, in other words, if the charge is decoupled.

2) The explosion pressures are one to two orders of magnitude below the elastic modulus E (static, in compression). E represents the pressure that should be applied to obtain 100% shortening of the specimen if it could behave elastically at any pressure (it certainly does not happen, due to both plasticization and breakage). In the layer close to the holes, the rock could continue to behave elastically only if strains of various units percent were accepted by remaining in the elastic range, which is certainly impossible. There is, therefore, no elastic return, and the holes remain "dilated" (when the "half casts" after the blast are noticeable, the holes undergo a measurable increase in diameter).

At this point, in the surrounding vicinity of the hole at the time of the explosion, apart from cases of low loading densities, the rock is closer to plasticity than to elasticity and, in the layer close to the charge, even "pasty".

3.3   The role of "plasticization"

The phenomena mentioned above pertain to that small portion of rock adjacent to the blast hole, often two or three times its diameter, and believed to be negligible: this is wrong, as they waste a noticeable fraction of the explosion energy. The work required to break a unit volume of "brittle" solid which keeps proportionality between the applied load and the strain up to the breaking point, is almost negligible: it can easily be shown that it is equal to $\sigma_{RC}^2/2E$ (σ_{RC} being the strength and E being the elastic modulus). But if the body yields plastically after breaking, the work increases by several times. Figure 3.4 clarifies the concept.

A cubic block (σ_{RC} = 100 MPa and E = 50,000 MPa) is loaded with an increasing force F, transmitted by a distribution plate. If it collapses abruptly, it absorbs about 100 kJ (sketch 1); if it is shortened by 10 mm before breaking, it has absorbed 1 MJ (sketch 2).

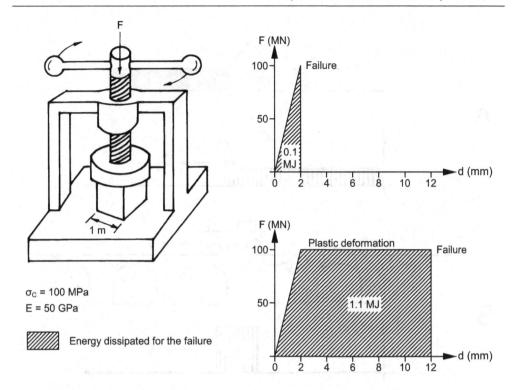

Figure 3.4 Influence of the plastic strain, before breaking, on the specific breaking work.

Obviously, in dynamic conditions E, σ_{RC}, and plastic strain cannot maintain the same values of the static tests, nor can the dynamic values be derived from static tests by simple multiplication by a corrective coefficient.

The example only wants to clarify the importance of a parameter that is sometimes overlooked in mechanical tests: the "strain at breakage". On the other hand, it is known that blasts in *weak* rocks often require a lot of explosives (e.g., gypsum), as some of the available work is spent in plastic deformation.

3.4 The role of "shock" and "quasi-static" actions

The explosion involves dynamic (lasting a few μs) and "quasi-static" pulses (lasting several milliseconds) into the medium (Anderson et al., 1983; Edri et al., 2011; Savir et al., 2009; Feldgun et al., 2016).

Both involve more or less the whole volume to be exploited, but the former doesn't act "simultaneously"; they rather "cross" it, at a speed of a few km/s, and therefore, the single points of the rock mass are affected for a very short time. Although their intensity is higher than the rock "strength", they do not cause a "complete" breakage due to their short duration, as the fractures require a certain time to develop (they propagate at a maximum speed of 1 km/s or less).

Impulsive stresses can mark the rock rather than break it like the diamond marks the slab, which then breaks under a weak effort. The "quasi-static" action, which lasts longer, completes the fragmentation (Gheibi et al., 2009; Siddiqui et al., 2009; Silva et al., 2016).

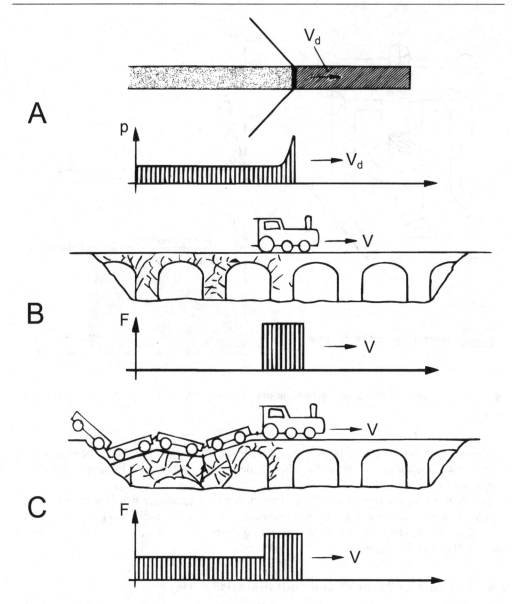

Figure 3.5 Cooperation between an impulsive load and a static load. A: Diagram of the detonation of a charge and pressure distribution on the wall of the hole at a given
instant. B: Effect of the impulsive load only on a bridge. C: Effect of the impulsive
load followed by a static or "quasi-static" load. *F*: load; *D*: detonation velocity; *p*:
Pressure; *V*: travel speed of the load (vehicle).

This way of thinking, one of the many possible ways, is shown in Figure 3.5.

The locomotive, while weighing more than the pillars, was able to reach the end of the bridge,
leaving it standing, though damaged, but passed over each pylon for a shorter time before col-
lapsing; however, if it is followed by a train, even though the wagons are lighter, it collapses,
because the load stays longer on each damaged pile.

3.5 Conversion of pressures into tensile strengths

Breakage by "crushing" only affects the rock close to the charge: most of the fragmentation from blasting is achieved by tensile failure (Grundstrom et al., 2001; Kansake et al., 2016; Hjelmberg, 1983).

In addition, it is worth mentioning that compression does not cause the breakage; rather, it is considered its "by-product".

The explosion pressure, applied to the wall of the blast hole, induces tensile stresses in the rock through two mechanisms, both involving the presence of "free surfaces" (boundary surfaces between the medium itself and another medium with practically zero strength: atmosphere, water, a heap of already fragmented material).

The first mechanism is the conversion of the compression pulse into a traction pulse by "reflection on the free surface" and obviously applies only to impulsive loads (Banadaki, 2010); the second is related to the so-called "pressure vessel model" and is applied for impulsive loads as well as for static or quasi-static loads: to better understand this concept, it may be useful to reflect on the simple test shown in Figure 3.6.

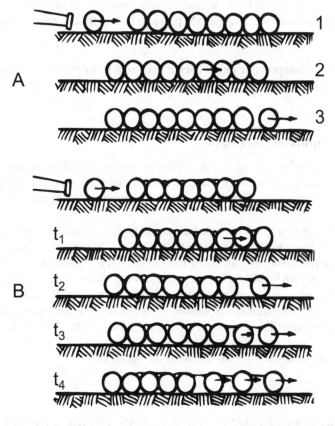

Figure 3.6 Example of reflection of a pressure pulse as a tensile pulse. A. The material has no tensile strength. 1: initial situation; 2: during the transmission of the pressure pulse; 3: final situation. B. A certain tensile strength is given to the material. t_1–t_4: situations in succeeding instants.

Returning to the billiard analogy, a one-dimensional phenomenon (balls aligned, rather than scattered on a plane) is studied for simplicity. The practical effect of the impact "reflection" is represented by parallel (or almost parallel) fractures to the free surface.

3.6 The concept of shock wave reflection on free surface

A: A row of connected balls, placed on the billiard table, undertakes a strike with another ball, colliding with the first ball of the row: (1) the pulse is transmitted from one ball to the next, and so on. If it were possible to pause the situation instantly after the first impact, it would have shown, for example, the ball in compression (imperceptibly "ovalized" by the impact received, which it is about to transmit to the next one); (2) the last ball reached by the pulse can't transmit it to another ball, so it moves with a speed (practically) equal to that of the first stricken ball; (3) the row of balls can transmit pressure, but not traction, so the last one cannot "drag" the previous one, etc.: the row remains standing.

B: The same test is repeated but, first, a "tensile strength" is exerted on the balls, simply by connecting the balls with a stretch of rope. At time t_1, immediately after the first collision, everything still happens as in **A**. When the pulse reaches the last ball (t_2), it moves, but immediately pulls the string, dragging the previous ball: at time t_3, this, in turn, drags the one before it, and so on.

It can be said that the compressive pulse, once it reaches the "free" ball, is reflected as a tensile pulse. If the rope is too weak at some points, it tears: this means that a "tensile failure" has occurred.

3.7 Tensile impulsive strength

The mechanism of "reflection" on a free surface or, in the one-dimensional case, on a free end, is used to measure the "dynamic" tensile strength, that is, a strength that lasts for a very short time. The experimental (simplified) scheme of the test (Hopkinson bar) is shown in Figure 3.7.

This way of determining the tensile strength has shown that the "dynamic" tensile failure requires a much higher stress (up to three times) than the "static" failure. Moreover, the static tensile strength on the metric scale (the common scale of interest in blasting) is yet a small fraction of the tensile strength, also static, shown by the decimetric specimens used in the laboratory tests. Furthermore, in most cases, the tensile strength on the metric scale is in fact very low (a "metric" volume of rock can be frequently affected by damages or weakening, which has little influence on the compressive strength but, practically, nullifies the tensile strength). Although the dynamic tensile strength is greater than the static one, it remains in practice very low in absolute terms: the mere "presence" of tensile stresses in a rock-mass means failure on the metric scale.

3.8 The "pressure vessel" model

By pressurizing a container, for example, a cylinder, it "swells" more or less visibly: its diameter (and therefore its circumference) increases. This means that the internal pressure induced a tensile stress on the wall. The force (F) that causes this stress can be easily calculated: , where p is the pressure, D the diameter, and l the height of the cylinder: it means that, if the wall of

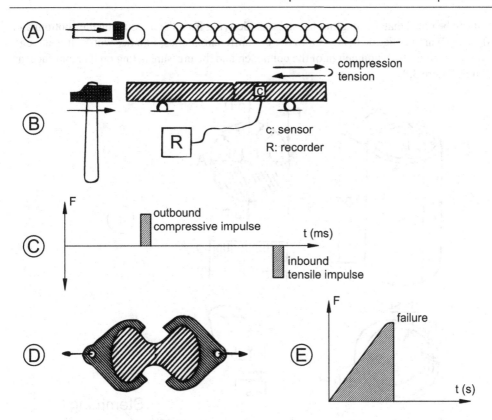

Figure 3.7 Dynamic and static assessments of the tensile strength of a material. A. Displaying the test principle. B. Scheme of the impulsive tensile strength test on a dented cylindrical rod. The breakage, if the pulse exceeds the tolerable strength of the material, occurs at the indentation. C. Stress/time diagram recorded by the sensor. D. Scheme of the static tensile strength test on an "eight" specimen (so-called dog bone). The breakage is noticeable in the narrowest section. E. Stress/time diagram in the static test.

the container was cut along a generator, the two edges should be kept close together with the force F to counteract the pressure p.

The tensile stresses are of course "circumferential": they tear the wall while trying to split it according to the generators. A blast-hole, upon explosion, can be considered a cylinder whose rock walls are filled with pressurized gas; the tensile stresses, when exceeding the rock strength, open radial fractures in the walls. This analogy is substantially true whether the impulsive or the "quasi-static" pressure is considered; it is also the oldest used for didactic purposes: in fact, at the time of black powder (negligible impact effects), it was the most reasonable explanation of the disruptive actions observed.

When applied to detonating charges, which provide a "shock wave" followed by a static action, the model is still valid; the two effects have just to be considered separately: the pulse stresses concentric *ideal* shells of rock and the static action keeps a tensile stress. For design purposes, it can also be postulated as a static action equivalent to the superimposition of the two.

The strength of the pressurized vessel can be easily studied if the vessel itself has a length much greater than its diameter (this generally happens in the case of blast holes). In this case, in fact, it can be noted that the longitudinal stresses (deriving from the pressure on the "bottom and on the "top") are negligible compared to the circumferential ones; in other words, it is enough to consider a ring of the unit height of the container, and the pressure acting on its inner face, as shown in Figure 3.8.

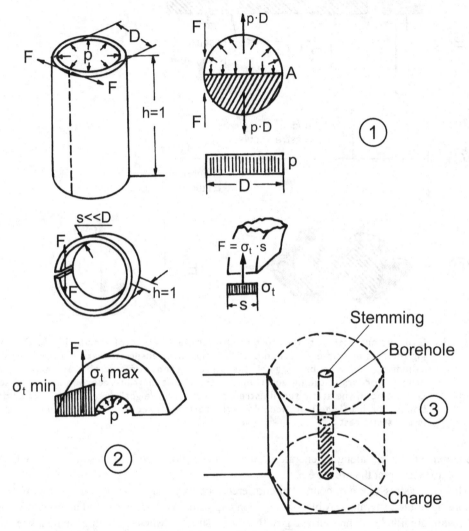

Figure 3.8 Schematic representation of the tensile failure on the walls of a cylindrical pressure vessel. D: diameter; p: pressure; F: tensile force; s: wall thickness of the vessel; σ_t: circumferential tensile stress.

1. Case of a thin-walled vessel, where the stress can be considered uniform in terms of thickness (Mariotte's formula). From the equilibrium of the moments with respect to A, it can be deduced: $F \cdot D = p \cdot D \cdot \dfrac{D}{2}$ from which: $F = p \cdot \dfrac{D}{2}$ by considering the traction uniformly distributed on the thickness s: $\sigma_t = \dfrac{F}{s} = p \cdot \dfrac{D}{2s}$ The failure occurs if σ_t exceeds the strength value.
2. Case of a thick-walled vessel: the stress along the thickness can't be considered uniform, as the stress σ_t is maximum on the inner walls and minimum on the external walls. The failure occurs if $\sigma_{t\,max}$ exceeds the strength value.
3. Example of a real case where the model of the pressure vessel can be applied.

When the blast holes are completely charged without decoupling, the local conversion of the rock to the "incoherent state" is very noticeable. The "quasi-static pressure" does not act on the "virgin" wall of the hole but on a shell of "rock paste", which, having no tensile strength, simply transfers the pressure to a coaxial cylindrical surface placed beyond the wall. Practically, everything happens as if the diameter of the hole was greater, and this relatively increases the destructive power of the charge. The effect can be guessed by considering the simple scheme of Figure 3.9.

In Figure 3.9, in 1, a cylindrical sheet metal vessel with diameter D_1 is pressurized with a pressure p. By means of the momentum equation with respect to A, the breakout force F_1 is obtained with the previously examined equation ($F_1 = p \, D_1/2$, in case of the unit height ring). F_1 is admitted being the maximum force that can be tolerated by that sheet. In 2, conversely, a vessel with diameter D_1 made by a plastic film (with negligible tensile strength) is pressurized with pressure p in a container made of the same sheet but with a diameter of $D_2 >$ D_1, and a practically incompressible fluid is placed between the two, water for instance: the pressure is transferred to the sheet, as the water reaches the pressure p, and the force F_2 that tends to tear the sheet becomes $F_2 = p \, D_2/2 > F_1$; the vessel yields. It is simply the principle of the "hydraulic press".

As soon as some fractures open on the wall of the hole, the pressurized gases penetrate and press on the walls to spread them apart. Figure 3.10 and its explanation are taken from the work of Christmann (1971). They are didactically effective, even though the wedging effect is controversial, at least in the early stage of failure under the action of the explosive.

However, the hypothesis leads, qualitatively, to the same conclusion as the considerations developed above on the postulated effect of plasticization: everything happens as if the "quasi-static" pressure acted on a cylindrical wall with a diameter greater than the original diameter of the hole.

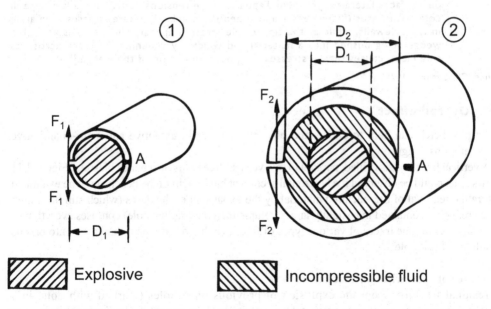

Figure 3.9 Scheme of the effect of an incoherent plasticized rock layer around the charge, after blasting. The mine is, for the sake of simplicity, considered a thin-walled vessel under pressure. 1: The charge is in direct contact with the rock. 2. The charge acts on an incompressible fluid between the charge itself and the rock. F: tensile forces; D: diameters; A: hinge points.

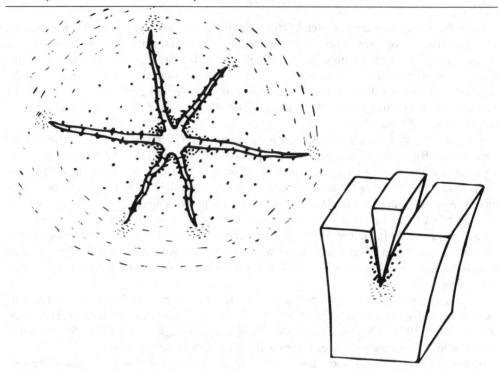

Figure 3.10 Gas wedging effect: "Les gaz (fumées) pénètrent dans les fissures qu'elles écartent. Des tensions de pression prennent alors naissance dans les flancs tandis qu'à la pointe se manifestent des tensions de traction. Cela se passe comme pour ce coin qu'on enfonce dans un matériau et qui, en écartant la fente par l'intérmediaire de ses faces latérales, provoque l'apparition de tensions de traction à la pointe di coin. [The gases (fumes) enter and widen the cracks. Pressure stresses then arise in the sidewalls, while at the tip, tensile stresses appear. This happens as with a wedge being pushed into a material and which, by widening the crack across its side faces, causes tensile stresses to appear on the tip of the wedge.]"

Source: Christmann (1971)

3.9 Overall effect

A simple sketch can help to understand what the provided qualitative and semi-quantitative considerations suggest.

A vertical blast hole is drilled, parallel to two vertical free walls, in a bench blasting. Figure 3.11 shows what could be observed if a horizontal section at half height of the bench could be examined before the removal of the fractured material by the expansion of the gases (which still have high pressure, being contained in a slightly larger volume than the original hole) confuses everything.

In this section, the traces of various types of fractures that divide the mined mass into prisms should be observable:

- an area of "pulverization", around the hole;
- residual fractures from the explosion of previous blast holes (marked with dots; in a systematic blast, a mine can never act on "virgin" rock, except at the beginning);
- parallel, or almost parallel, fractures to the free surfaces, due to the "reflection" effect, more frequently close to the free surfaces, not present behind the hole;

- radial fractures, converging on the axis of the hole and more densely distributed in its vicinity, that develop through the free surfaces more than towards the rock mass which will remain in place at the end of the blast.

3.10   The noticeable effect of the explosion

The rough sketch in Figure 3.12 depicts a reasonable theoretical reconstruction of an intermediate stage: what can be observed is the final result of the blast, if it works correctly: the heap of a

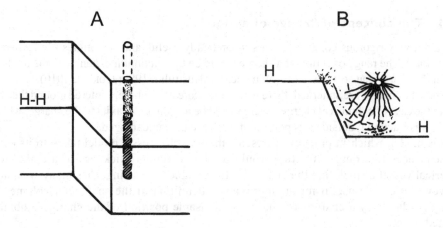

Figure 3.11 Ideal distribution of the fractures induced in the rock-mass by the explosive in bench blasting. A. Perspective view of the bench. B. Hypothetical horizontal section, immediately after the blast.

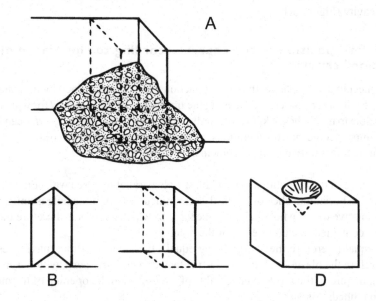

Figure 3.12 A. Cavity created by the detonation of a blast and "muck-pile" obtained. The amassed shape of the heap shows that the operation was carried out correctly. B, C, and D. The geometry of the cavities is produced by a blast hole with, respectively, two, three, and one free surface.

fragmented material, of about 1.5 times the volume of the original rock in place, from which the granulometric distribution and the volume of the residual cavity can be determined. The latter, in the case of cylindrical charges, has a roughly prismatic shape, with an edge corresponding to the axis of the blast hole.

If there is only one free wall, the removed volume is a prism with a triangular base, while in the case of two free walls, the base is a parallelogram, as shown in Figure 3.12.

When the charges are spherical ($L \leq 6\,\Phi$) and are blasted with a convenient burden, the charge is the vertex of an open conical cavity, which corresponds to the volume removed (crater).

3.11 The concept of "range of action"

It is a not very rigorous concept, but it's theoretically useful, and sometimes used when planning a blast. The range of action of a charge would be the maximum distance, in the medium, at which the charge can still exert a breaking action (Hustrulid, 1999; Gokhal, 2010).

When the charges are spherical, there would therefore be a sphere, centered around the charge, and in the case of cylindrical charges, there would be a cylinder, coaxial to the charge, where the medium is fractured; outside this portion, it would remain undamaged.

This model, which in practice represents the pressure vessel model taken to its extreme consequences (the range of action would be the maximum thickness of a cylindrical or spherical vessel of rock that the charge is still capable of blasting), has many weaknesses. However, it can be useful for an approximate definition of the shape and volume of the cavity produced by a charge, and the most advisable position of the charge to obtain the wanted effect.

To provide an observable effect, of course, the minimum distance of the charge from a free wall (the so-called "burden") must be less than the "range of action". The intersection of the free wall with the "cylinder" or "sphere" must represent the contour, on the free wall itself, of the parting cavity (Figure 3.13).

3.12 Optimal position of the charge according to the "range of action" criterion

Although conceptually debatable, the range of action r can be determined by trial and error, for a certain charge with a certain rock, by seeking the limit value of the burden d for which that charge is no longer able to break the rock, even partially. Under such conditions, $r = d$. r can therefore be used, with simple geometric considerations, to find the optimum position of the charges.

The example refers to two simple problems:

1) Optimal value of the r/d ratio: a vertical blast-hole parallel to a free wall (Figure 3.14) is still considered; a horizontal slice of height 1 is examined, which, if the hole is uniformly loaded, is representative of the whole height; the charge to which r refers is therefore the amount of explosive contained in a unit height of the hole.

   Being r equal (therefore the charge being equal), the value of α that maximizes the area S (and therefore the blasted volume) is $45°$ (in fact, $S = r^2.\ sen\alpha.\ cos\alpha$, which is maximum for $\alpha = 45°$); therefore, the most suitable value of r/d is $\sqrt{2}$ and the opening of the most efficient separation dihedral is $90°$.

2) Optimal ratio between the spacing and the burden: in a row of blast holes that detonate according to a given delay sequence in a vertical bench blasting.

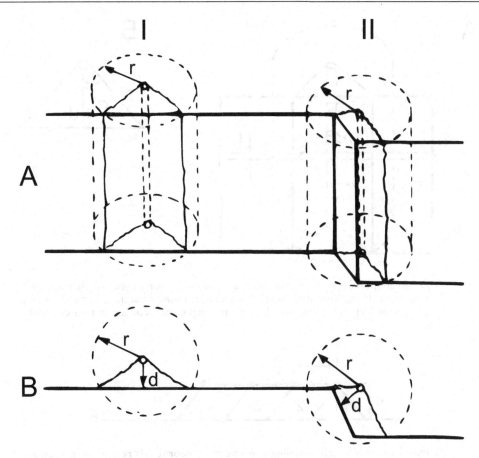

Figure 3.13 Concept of "range of action" of a cylindrical charge. I. Blast-hole with two free
surfaces. II. Blast-hole with three free surfaces. A: perspective view; B: plan view;
d: burden; *r*: range of action.

Figure 3.15 shows the horizontal projections of the subsequently blasted volumes. With the
same simple logic adopted for a single blast hole, it can be observed that the most effective
value of i is given by $i = r$, and therefore, the optimal value of i/d is $\sqrt{2} = 1.42$. Anyway,
a:.2 ratio is currently adopted, which is lower, to be more certain of obtaining a regular
residual wall.

3.13   The comminution effect

In heavily fractured rocks, excluding the immediate vicinity of the charge, where pulverization
can occur, the explosive only enlarges the existing fractures and develops the latent ones. The
size of the blasted material practically corresponds to that of natural monoliths, and it would be
improper to speak of "comminution".

In sound rocks, the distribution of the fractures induced by the explosion according to the described
scheme, followed by the disconnection of the elements isolated from the fractures, can be considered

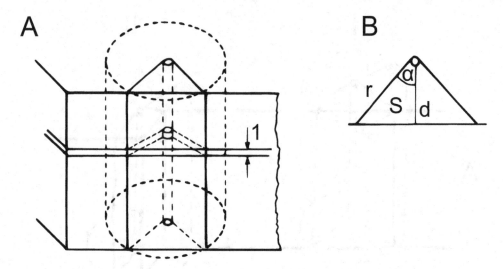

Figure 3.14 Optimal opening of the prism, according to the action range principle. A: perspective view; B: horizontal section; *d*: burden; *r*: range of action; *a*: opening angle of the prism. When looking for the optimal angle, the problem is two dimensional.

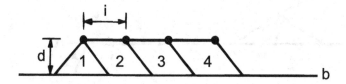

Figure 3.15 Plan view of a bench, showing the optimal (theoretical) relationship between the spacing and the burden, according to the "range of action" principle. *d*: burden; *i*: spacing; *b*: bench face; 1, 2, 3, 4: detonation sequence.

the effect of a comminution like that taking place in a primary crusher (Figure 3.16). It may be useful to consider the work of the explosive from this point of view, transferring some concepts of crushing technology towards blasting: like the crusher, the blasting system can also be "adjusted" within certain limits (although in different ways) to give a finely comminuted "product".

The concepts briefly examined below are

- the particle size distribution;
- the reduction ratio;
- the comminution laws;
- the specific comminution work.

3.14   Particle size distribution and characteristics

In a muck pile, the individual fragments have different sizes; to evaluate their sizes and establish which material is thicker or finer than the other, or how much thicker or finer it is, a comparison between the individual fragments is not enough: instead, the "particle

A

B

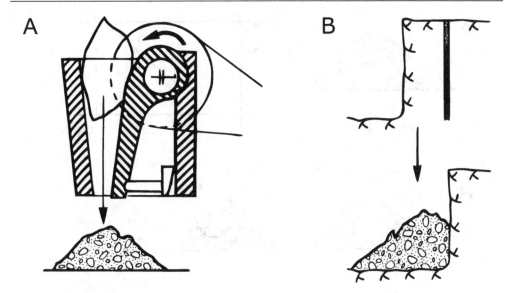

Figure 3.16 Analogy between a crushing machine (A) and a blast-hole (B). Intuitively, in the first case, a piece of rock is pressed between a fixed and a mobile plate; in the second, the charge plays the role of the mobile plate and the inertia of the rock that of the fixed plate.

size distributions" of the two muck piles must be compared: curves (or tables) showing the percentages of large, medium, and fine sizes (Kuznetsov, 1973; Scott et al., 1998; Onederra et al., 2004).

The ways of determining the results are different: one of them is shown (the passing %/size curve) in Figure 3.17 for conceptual clarification (obviously, when analyzing the grain size distribution from a blast, tools other than sieving are used). After the analysis of the two I and II heaps has been done, two diagrams are obtained. The comparison immediately shows that the distribution *I* is finer than *II*. However, it can sometimes be inconvenient to do calculations and comparisons on diagrams or tables. By trying to "condense" the diagram into a single number (although this is questionable and arbitrary), a conventional "characteristic dimension" can be taken as an indicator of the whole distribution. In the example, d_{50} was adopted (the size of the sieve openings that allows 50% of the material to pass), but other conventions are also possible (and used), such as d_{10}, d_{75}, and d_{90}; it is obviously important not to change the convention throughout the process.

Besides the "characteristic dimension", another parameter is also necessary to indicate the intensity of comminution: it is called the "reduction ratio" (R). As for the crusher, the input material has a given characteristic dimension (d_1) greater than the output material (d_2), and it is natural to use the ratio d_1/d_2 as an indicator of the intensity of comminution, which is the "reduction ratio": by reducing the outlet opening to further reduce the size of the fragments, d_2 decreases and therefore R increases; the opposite occurs if the outlet opening is enlarged. The example of Figure 3.18 clarifies the concept.

By scaling up the concept to the case of blasting, there is a conceptual difficulty, as the material, at the beginning, does not have a "characteristic dimension": however, being a conventional value, its role can be assigned to a geometric dimension characterizing the excavation

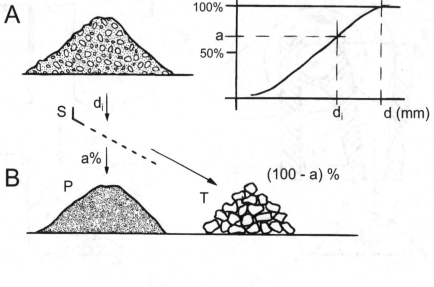

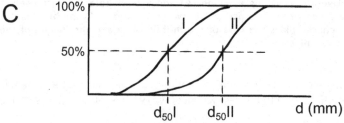

Figure 3.17 Concept of "particle size distribution" (A), of "cumulative class" (B), and of "characteristic size" (C) of a fragmented material.

In A, the heap, and its grain size distribution curve (on the horizontal axis, the dimensions of the materials, on the vertical axis the cumulative percentage of passing through).

In B, one of the ways to determine a point of the distribution curve is shown: by means of a metal grid S with an opening d_i, a "passing" or "under" P is obtained, of which the percentage "a" with respect to the total can be easily calculated, and a "retained" or "over" T, whose percentage is also calculated. In this way, the "a/d_i" point of the curve is obtained. The same procedure, applied with grids of different openings, allows to determine the other points.

In C, two materials, one finer (I) and one coarser (II) are fully described by the relative particle size distributions. For a less accurate but more convenient description, the "characteristic dimensions" d_{50I} e d_{50II} can be used.

system; the simplest choice is to use the "burden" (this choice is also conceptually correct: a blast hole with a minimum, but still effective, charge, must at least detach some monoliths of that size). Also for a blast, a "reduction ratio" can be then defined, characterizing the fragmentation obtained as shown in Figure 3.19.

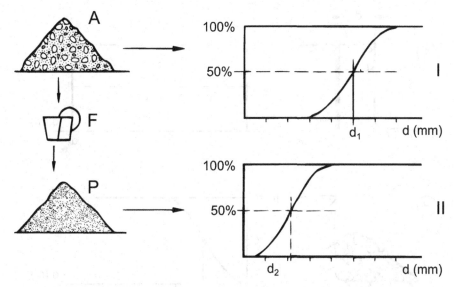

Figure 3.18 Concept of "reduction ratio" in a crushing operation. An "input" material (A) is characterized by its granulometric distribution curve (I) or, more briefly, by its characteristic dimension d_1. By crushing the material through a crusher F, a product P is obtained, characterized by a different grain size distribution curve (II) or, more briefly, by its new characteristic dimension d_2. Obviously, $d_2 < d_1$; the reduction ratio is d_1/d_2.

3.15 Comminution laws

Comminution laws are (theoretical) relations that link the specific comminution work (J/m^3) and the "reduction ratio".

It is intuitive that, for example, the work required to reduce a volume of 1 m^3 into fragments with a side of 30 cm (i.e., with a reduction ratio of about 3) is less than that required to reduce it into fragments with a side of 3 cm (i.e., with a reduction ratio of about 30).

Two extreme "laws" were conjectured:

- *Rittinger law*: the specific work L necessary to attain the comminution of a solid is a function of the reduction ratio R, according to the law: $L = Cost. (R-1)/d_1$, where *Cost.* is a constant expressed in J/m^2 characteristic of the medium, and d_1 is the input characteristic size.
- *Kick law*: the specific work L necessary to attain the comminution of a solid is a function of the reduction ratio R, according to the law $L = Cost. \log R$, where *Cost.* is a constant expressed in J/m^3 characteristic of the medium (notice that in this "law", the input characteristic size d_1 does not appear).

Both laws seem theoretically well reasoned but give different results when applied to the same case, so it is common to quote the "Kick-Rittinger paradox"; in practice, intermediate, compromise "laws" are used in the comminution technique (Radziszewski, 2013; Austin, 1973; Stamboliadis, 2013; Rumpf, 1973).

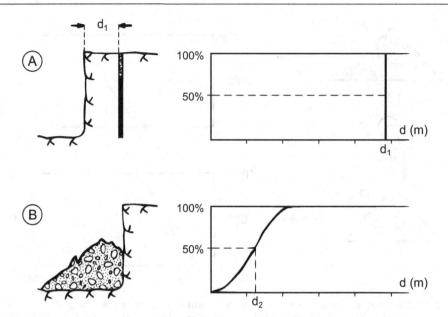

Figure 3.19 Extension of the concept of "reduction ratio" to the excavation by blasting.

> To its initial state (A), the rock is on-site, not blasted yet but already charged; it can be assigned a "characteristic dimension" d_1 equal to the burden or, in case of timing between a row of blast holes, to the blasting pattern.

> After blasting (B), the original volume has been reduced to a muck pile of fragmented material, of which the "characteristic dimension" d_2, obviously lower than d_1, can be determined. The ratio d_1/d_2 is the reduction ratio.

A comparative analysis of the two "laws", starting from their theoretical assumptions (fragmentation work proportional to the fracture surface, according to Rittinger; fragmentation work proportional to the volume to be broken, according to Kick) would show that the former is to be preferred when the comminution occurs mainly due to shear failures, the second when it occurs mainly due to tensile failures (Vesilind, 1980; Ouchterlony, 2003; Sanchidrián, 2010; Ouchterlony and Sanchidrián, 2019). The case of blasting and that of coarse crushing would be closer to Kick's approach, that of fine grinding to Rittinger's reasoning. However, no theoretical law, whatever it may be, provides a reliable model of the real phenomenon.

It is a lucky circumstance that the comminution law of blasting is closer to Kick's model than to Rittinger's: it is not necessary to brutally increase the charge to increase R, meaning that for the same blast holes' geometry, to reduce the grain size distribution. However, it is always a theoretical assumption. Figure 3.20 clarifies the divergence of the two laws with a simple example.

As for the mechanical strength of rocks, the theoretical laws of comminution have been above all discussed to induce mistrust in theoretical calculations: they give valid results only if, regardless of the intrinsic goodness of the formula, they are calibrated with coefficients and experimental constants, which can only be derived by processing the data collected from real cases.

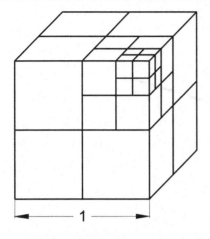

Cube division	Reduction ratio R	Comminution work (arbitrary units)	
		KICK	RITTINGER
8	2	1	1
64	4	2	3
210	8	3	7
4096	16	4	15

Figure 3.20 Explanation of the divergence between the two comminution laws (Kick and Rittinger) for the prediction of the specific comminution work of the wanted reduction ratio. The reduction into finer fragments of a cubic unit volume of a hypothetical material is considered.

3.16 Specific comminution work

The specific comminution work consists in the actual unit energy cost (i.e., referred to a unit volume of material), and in the case of crushing, it is logically expressed in kWh/m^3; in the case of blasting, it can be more logically expressed in kg of explosive/m^3, although it depends both on the material and on R.

In both cases, the theoretical laws of comminution provide that the energy cost is zero when R is 1, but this is certainly false. Referring to the crusher, although it has the output opening equal to the input dimension of the material, that is with zero effect ($R = 1$), a certain amount of power would still be absorbed by the "idle running" of the machine (and it is not small: in many crushers it is more than 1/3 of the power absorbed in the common production running). Similarly, referring to blasting, there is a minimum charge below which there is no effect ($R = 1$), and this charge is generally not negligible.

By plotting the unit energy cost as a function of R, a curve that passes through the point 1–0 can't be found, but something like the graph in Figure 3.21: only the excess of work with respect to a minimum value seems to be exploitable.

Curves of this type can be obtained by examining the results of systematic blasts; they could also be used to "adjust" the size of the product, but in this regard, as mentioned, there is a big difference between a blast hole and a crusher.

At this point, the analogy between the crusher and the blast ends. In fact, the crusher only provides one chance of adjustment, by narrowing or widening the outlet opening: with the first option, the flow rate is reduced, as the time the material is forced to remain in the machine increases (since the volume of the comminution chamber undergoes a slight variation) and, consequently, a greater comminution work is provided to the unit volume of material; by widening the outlet opening, the opposite effect occurs.

In the case of blasting, instead, there are two possible settings: the powder factor (kg/m^3) can be increased or reduced, or the pattern can be adjusted. In the latter case, according to the convention

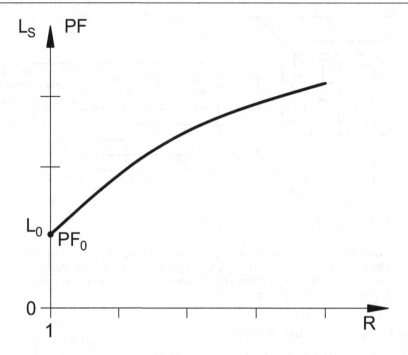

Figure 3.21 Trend, on an arbitrary scale, of the relationship between the specific work of comminution (or its equivalent, which is the specific consumption of explosive) and the reduction ratio obtained. L_s: specific comminution work, J/m³; P.F.: specific consumption of explosive, kg m³; L_0: work required by the no-load running (with $R = 1$) of the commutation machine; P.F.$_0$: minimum effective specific consumption of explosive (specific charge below which there is no effect); R: reduction ratio.

here adopted, the "characteristic dimension" d_1 is considered, not necessarily R, but also the "characteristic dimension of the product" d_2 can be modified, according to the needs of the specific case.

Figure 3.22 clarifies the meaning of the two types of regulation.

The choice of one type or the other depends on the ratio between the unit costs of drilling and charging, and on technical limitations or rules imposed for different reasons.

3.17 Cases where the comminution is not wanted (splitting)

In many cases, the goal is to obtain a single fracture to isolate a given volume of rock without damaging it or producing other cracks.

When this is the goal, impulsive actions are reduced or abolished, and the "quasi-static pressure" is reduced, according to different techniques: the most common consists of "decoupling" the charge (using cylindrical charges with a diameter smaller than that of the hole), which also leads to a reduction in the charging density (ratio of the mass of the charge to the volume of the hole). The holes are aligned exactly according to the wanted splitting surface and the blast is strictly simultaneous (Hino, 1956).

In these cases, the explosive practically works as a pressurizing agent, with an effect like static pressure; the pulverized band around the holes almost disappears. The "guidance" of the fracture becomes important, both in isotropic and, above all, in anisotropic rock (where it is not always successful).

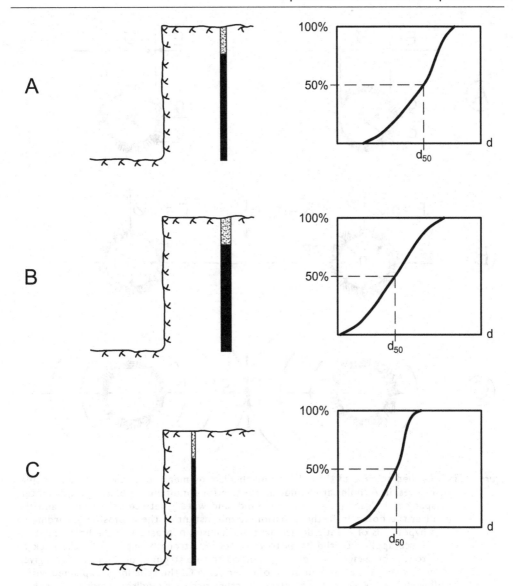

Figure 3.22 Different possibilities of adjusting the fragmentation in a blast, by varying the charge or the blasting pattern.

In A, a reference scheme is characterized by a given specific charge, a given burden, and a certain grain size distribution of the blast. In B, to increase the fragmentation, the specific charge was increased, keeping the burden unchanged; in C, the same effect was sought by maintaining the same specific charge and reducing the burden.

Note that the two changes can give the same result in terms of D_{50}, but not necessarily in terms of the whole grain size distribution, as the graphs show.

The simplified mechanism of action of the charges is described in Figure 3.23. Considering a row of charged holes and a slice, normal to the axis of the holes, of unit thickness, the failure condition is examined according to the static mechanism: F is the force exerted by the gases, R is the resisting force, p is the explosion pressure at the wanted charging density, l is the distance

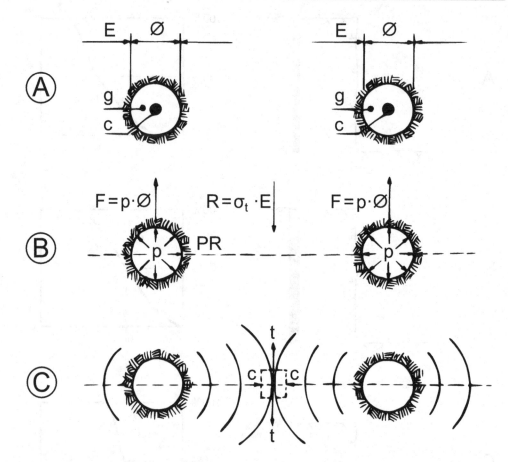

Figure 3.23 Simplified scheme of the splitting mechanism by means of a row of parallel coplanar holes. In A, holes are loaded and set up for simultaneous blasting; d: diameter;
l: spacing; g: "gap" (filled with air, sand, and water) between the charge and the
wall of the holes. In B, the situation at the instant of the explosion according to
the hypothesis of a static detachment mechanism; p: pressure in the hole; F: resultant force applied by the holes to the rock; R: strength opposed by the rock to
the tensile breakage; PR: splitting plan; σ_r: tensile strength of the rock, for a given
speed of application and duration of the load. In C, the splitting is explained with a
different theory, whereby the stress fracture occurs by indirect traction (t) generated by the intense compression C due to the impact of the shock waves coming
from the blast holes.

between the holes, and σ_r is the tensile strength at the application velocity and duration of the
load considered; in such conditions:

$F = R$, that is: $p \times D = \sigma_r l$

Comments and explanations

The "pressure vessel" model was used.

The most important parameter is obviously the ratio of the spacing of the holes to their diameter: in fact, $Pd > \sigma_R l$ can be rewritten as $p/\sigma_R > l/d$, and p/ρ_R, for a given explosive/decoupling/rock, can be considered a constant.

The concept of "specific consumption" (powder factor) is irrelevant in designing blasts of this type; in fact, balances of energies are not considered (explosion work/fragmentation work) but of forces (active force/resistant force).

Data referred to the surfaces, more than those referred to volumes, are useful, that is, the "active surface", that is, the product of the hole's diameter by its charged length, and the "resistant surface", the product of the spacing of the holes by their length. To succeed, adequate relationships between these areas must be respected, as opposed to the volumes as in conventional blasts.

For this type of work, it is not the same (as a result and as a cost) to use holes with a diameter of 30 mm spaced by 300 mm or holes with a 100 mm diameter spaced by 1 m: it can be easily calculated that the specific consumption of explosive referred to the splitting surface (g/m^2) decreases with the decrease in the diameter of the holes (ICMM, 2012).

The "gap" between the explosive and the hole wall can be filled with air, water, or sand: in the first case, a greater charge is required for the same splitting surface; filling with water, where possible, is the most practical and effective way.

Blasting is preferable to be simultaneous; where this is not possible, it can be micro-delayed. The use of long delays does not give good results.

A single blast hole initiated at an infinite distance from a free wall results in random radial fractures; by reducing the pressure (e.g., by reducing the charge density), at a certain point the explosion no longer even gives this observable effect: the wall of the hole was stressed, it expanded slightly, then returned to its original condition (when pressure stops). In Figure 3.24, these situations are observed in diagrams A and B. The same happens, of course, if in the same medium, several blast holes detonate very far from each other, even if parallel and coplanar (for "very far from each other", it is meant "at a distance much greater than their range of action").

By progressively reducing the mutual distance between the holes, in the random ray of fractures produced by each of them, two fractures can be distinguished for each blast hole, that interconnects with the two produced by the adjacent holes; a more or less regular surface is obtained which interrupts the mechanical continuity of the medium, as well as irregular cracks affecting a strip of rock adjacent to this surface (situation C). By also reducing the pressure, a single fracture (actually, an alignment of fractures connecting the holes) can be produced; in this case, the fracture is oriented along the row of holes (situation D).

In line with "quasi-static" terms (an approach that seems preferable at low charge densities and at high decoupling), several conditions based on elementary considerations must be respected:

1) The distance of the holes from the free wall must be large enough (in particular, it must be greater than the distance between the holes) to avoid some fractures prefer to develop towards the free wall.
2) The tensile stress induced radially by the pressure of the gases around the single hole must be, in terms of intensity and duration, lower than that necessary to open the fractures (otherwise random fractures would develop).
3) On the contrary, in the superposition plane (the plane where the holes lie), the stresses induced by the pressure acting in the adjacent holes must be sufficient to induce the failure, as shown in Figure 3.25.

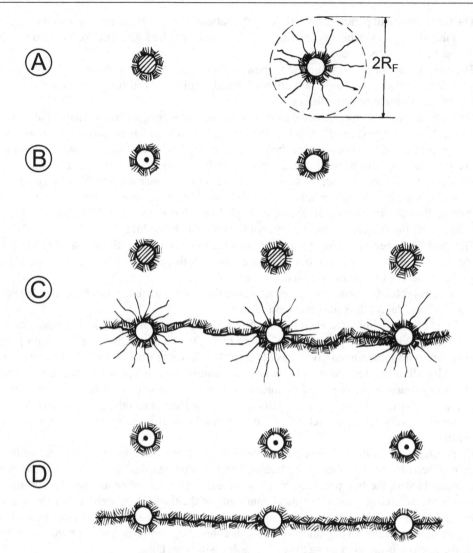

Figure 3.24 Schematic diagram of the splitting effect, in case of fully coupled (A, C) or decoupled (B, D) holes.

A. A hole with a fully coupled charge is surrounded, after the explosion, by a ray of fractures that extend up to a certain RF distance (fracturing radius). If the free wall, or another hole, is placed at a great distance, there is no other effect on the rock.

B. Under the same conditions, a hole with a decoupled charge does not give rise, after explosion, to any fracture.

C. An alignment of coupled charged holes, suitably close, results in a fracture where the network of radial fractures of the individual holes is superimposed.

D. Under the same conditions, an alignment of holes with decoupled charge only results in the splitting fracture.

This "static" point of view, certainly questionable and inaccurate, accounts for the technique currently adopted to obtain the guidance, which does not consist in falling below a certain decoupling value (usually 2), but by keeping above a certain limit the ratio of the diameter of the holes to the spacing (usual range from 1/6 to 1/12).

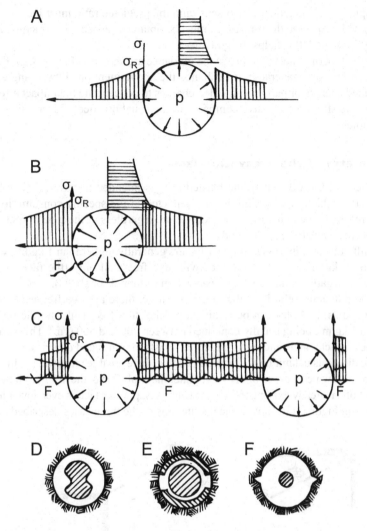

Figure 3.25 Simplified scheme of the guidance mechanism of a fracture along an alignment of blast holes.

A. A single hole is compared to a pressure vessel with infinite-thickness walls. In the walls, the internal pressure P gives rise to circumferential tensile stresses σ, which decrease as the distance from the wall increases; if σ does not exceed a certain limit σ_R, there is no failure.

B. Again, in the case of a single hole, if P increases, the circumferential stress σ also increases and, if it exceeds σ_R, a failure occurs (F); however, due to the symmetry of the system, it can start randomly from any point on the wall.

C. If more closely spaced holes are drilled and pressurized (as in A), the circumferential stresses overlap in the plane of the holes and, if the total exceeds σ_R, the failure occurs. However, the overlap is most effective in the plane where the holes lie, and this, therefore, "guides" the fracture.

In D, E, and F, experimental devices to improve the guidance are schematized. D: the charge is shaped to concentrate the impact on two generators; E: the charge is partially shielded by two metal half-shells ("charge holder"); F: two slits were made along two opposite generators of the hole wall.

In anisotropic rocks, the guidance worsens (and the predicted ratio must increase) if a transverse fracture with respect to the schistosity must be obtained (instead, if a fracture according to the schistosity is wanted, this helps the guidance).

Methods (still experimental) to improve the guidance are shown in Figure 3.25: the slit of the holes along two opposite generators (to enhance the above-mentioned "wedging effect"), and the use of shaped charges or partially shielded charges to concentrate the impact effect laterally. However, they are still too expensive compared to the usual practice of decreasing the spacing between the holes.

3.18   Geometry of the excavation face

The face is the wall towards which the blasted rock is intended to move. A first discrepancy among the works, which is very significantly reflected in the specific consumption of explosives, and therefore on the unit cost of the excavation, regardless of the type of rock, pertains to the number of free walls (Olofsson, 1988).

The most difficult case, involving powder factors generally higher than 1 kg/m³, with a maximum of up to 10 kg/m³, is that of the face with one free wall only (the face itself), that is, bounded only by negative dihedrals (e.g., excavation of tunnels or shafts).

If one of the dihedrals delimiting the face is positive, there are two free walls; the work is greatly facilitated as the holes can be arranged parallel to a free wall, and the powder factor hardly reaches 1 kg/m³, being usually contained between 200 and 500 g/m³. This situation is the rule in trenches and channels and in "production" blasts in mines and quarries.

If two of the dihedrals delimiting the face are positive, there are three free walls. This is generally the situation towards which a bench blast evolves after the explosion of the first blast hole of a row, or an enlargement of an already existing trench; in such cases, powder factors even lower are possible.

The sketches in Figure 3.26 show typical situations of the three cases described.

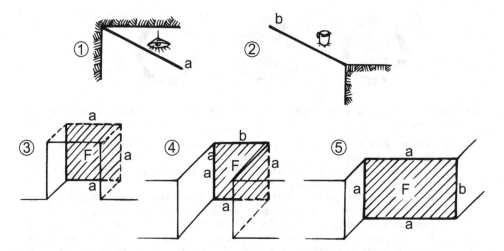

Figure 3.26 Schematic illustration of the excavation face with respect to the number of free walls.

1. Negative dihedral (*a*);
2. Positive dihedral (*b*);
3. Face (*F*) with a single free wall (blind excavation);
4. Face with two free walls;
5. Face with three free walls.

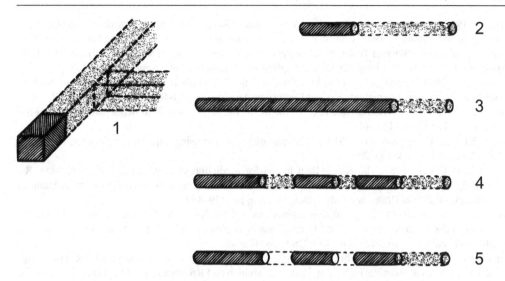

Figure 3.27 Examples of charge geometries.
1. Spherical charge (chamber blasting technology).
2. Spherical charge (cylindrical charge, but with small elongation).
3. Continuous cylindrical charge.
4. "Decked" cylindrical charges, with intermediate stemming.
5. Fractioned cylindrical charge with "air decks" (empty spaces between charges).

As for the geometry, the charges can be conventionally cylindrical (length/diameter > 6) or spherical (length/diameter < 6); see Figure 3.27.

In fact, the "cylindricity" or "sphericity" of the charge is not so relevant, but rather an opportunity to illustrate the "action volume" with a cylinder or a sphere. For example, an alignment of spherical charges slightly spaced along a hole is comparable to a cylindrical charge.

The cylindrical charges can be further subdivided into continuous and "decked", that is, charges separated by the interposition of stemming (in this case, it is necessary, in some way, to ensure the detonation continuity of the system); this material can also be air (air decking, used to get an effect similar to the above-mentioned decoupling).

Terms such as ordinary, small, or large blast holes are also commonly used: their meaning, of course, depends on the context; there are no accepted conventions on the size limits of "large" or "small" blast holes: it makes more sense to refer to small-, medium-, or large-diameter blast holes, because these terms can be matched to different types of drilling machines).

References

Afum, B.O. and Temeng, V.A., 2015. Reducing drill and blast cost through blast optimisation – A case study. Ghana Mining Journal, 15(2), pp. 50–57.

Anderson, C.E., Baker, W.E., Wauters, D.K. and Morris, B.L., 1983. Quasi-static pressure, duration, and impulse for explosions (e.g. HE) in structures. International Journal of Mechanical Sciences, 25(6), pp. 455–464.

Austin, L.G., 1973. A commentary on the Kick, Bond and Rittinger laws of grinding. Powder Technology, 7, pp. 315–317. Elsevler Sequoia S A, Lausanne – Printed in the Netherlands.

Banadaki, M.M.D., 2010. Stress-Wave Induced Fracture in Rock Due to Explosive Action. PhD thesis, Department of Civil Engineering, University of Toronto, Toronto, Canada, 284 pp.

Bozic, B. 1998. Control of fragmentation by blasting. Rudarsko-geoloiko-nafini zbornik, 10, pp. 49–57.

Calder, P.N. and Tuomi, J., 1980. Control blasting at sherman mine. Proceedings of the 6th Annual Conf. on 'Explosives and Blasting Technique', Society of Explosives Engineers, Montville, OH, pp. 312–320.

Chiappetta, R.F., 2001. The importance of pre-splitting and field controls to maintain stable high walls, eliminate coal damage and over break. Proceedings of the 10th High-Tech Seminar on 'State of the Art, Blasting Technology, Instrumentation and Explosives Application', GI-48, Nashville, TN, pp. 22–26.

Christmann, W., 1971. Le méchanisme tu tir au rochet. Revue de L'Industrie Minerale-Mines, Special issue, November 15, pp. 134–145.

Chung, S.H. and Katsabanis, P.D., 2010. Fragmentation prediction using improved engineering formulae. Fragblast, 2000(3), pp. 198–207.

Cunningham, C.V.B., 1983. The Kuz – Ram model for prediction of fragmentation from blasting. R. Holmberg and A. Rustab (Eds.), Proceedings of First International Symposium on Rock Fragmentation by Blasting. Balkema Rotterdam Pub., Luleå, Sweden, pp. 439–454.

Cunningham, C.V.B., 1987. Fragmentation estimations and the Kuz – Ram model – four years on. W. Fourney (Ed.), Proceedings of Second International Symposium on Rock Fragmentation by Blasting. Balkema Rotterdam Pub., Keystone, CO, USA, pp. 475–487.

Cunningham, C.V.B., 2000. The effect of timing precision on control of blasting effects. R. Holmberg (Ed.), Explosives & Blasting Technique. Balkema, Munich and Rotterdam, pp. 123–128.

Cunningham, C.V.B., 2005. The Kuz-Ram fragmentation model – 20 years on. R. Holmberg et al. (Eds.), Conference Proceedings 2005. European Federation of Explosives Engineers, Brighton, UK, pp. 201–210.

Dindarloo, S., Askarnejad, N. and Ataei, M., 2015. Design of controlled blasting (pre-splitting) in Golegohar iron ore mine, Iran. Transactions of the Institution of Mining and Metallurgy, Section A: Mining Technology, 124(1), pp. 64–68.

Edri, I., Savir, Z., Feldgun, V.R., Karinski, Y.S. and Yankelevsky, D.Z., 2011. On blast pressure analysis due to a partially confined explosion: I. Experimental studies. International Journal of Protective Structures, 2(1), pp. 1–20.

Esena, S., Onederra, I. and Bilgin, H.A., 2003. Modelling the size of the crushed zone around a blasthole. International Journal of Rock Mechanics & Mining Sciences, 40(2003), pp. 485–495.

Feldgun, V.R., Karinski, Y.S., Edri, I. and Yankelevsky, D.Z., 2016. Prediction of the quasi-static pressure in confined and partially confined explosions and its application to blast response simulation of flexible structures. International Journal of Impact Engineering, 90(15), pp. 46–60.

Fourney, W.L., 1993. Mechanisms of rock fragmentation by blasting. Comprehensive Rock Engineering Principles, Practice and Projects, vol. 4. Pergamon Press, Oxford, pp. 39–69.

Gheibi, S., Aghababaei, H., Hoseinie, S.H. and Pourrahimian, Y., 2009. Modified Kuz – Ram fragmentation model and its use at the sungun copper mine. International Journal of Rock Mechanics & Mining Sciences, 46, pp. 967–973.

Gokhal, V.B., 2010. Rotary Drilling and Blasting in Large Surface Mines. CRC Press, Boca Raton, FL, USA, 748 pp.

Grundstrom, C., Kanchibotla, S.S., Jankovich, A. and Thornton, D., 2001. Blast fragmentation for maximising the sag mill throughput at Porgera Gold Mine. Proceedings of the 27th Annual Conference on Explosives and Blasting Technique, vol. 1. ISEE, Orlando, FL, USA, pp. 383–399.

Hino, K., 1956. Fragmentation of rock through blasting and shock wave theory of blasting. Quarterly of the Colorado School of Mines, 51(3), pp. 191–209.

Hjelmberg, H., 1983. Some ideas in how to improve calculations of the fragment size distribution in bench blasting. Proceedings of the 1st Int. Symposium on Rock Fragmentation by Blasting, Luleå University of Technology, Luleå, pp. 469–494.

Hustrulid, W.A., 1999. Blasting Principles for Open Pit Mining. A.A. Balkema, Rotterdam, Brookfield Corporation, Canada, p. 1036.

ICMM, 2012. Trends in the mining and metals industry. Mining's Contribution to Sustainable Development, International Council on Mining and Metals, London, UK, 16 pp.

Jung, H.S., Jung, K.S., Mun, H.N., Chun, B.S. and Park, D.H., 2011. A study on the vibration propagation characteristics of controlled blasting methods and explosives in tunnelling. Journal of the Korean Geoenvironmental Society, 12(2), pp. 5–14.

Kanchibotla, S.S., Valery, W. and Morrell, S., 1999. Modelling fines in blast fragmentation and its impact on crushing and grinding. Proceedings of Explo'99 – A Conference on Rock Breaking, The Australasian Institute of Mining and Metallurgy, Kalgoorlie, Australia, pp. 137–144.

Kansake, B.A., Temeng, V.A. and Afum, B.O., 2016. Comparative analysis of rock fragmentation models – A case study. 4th UMaT Biennial International Mining and Mineral Conference, University of Mines and Technology Tarkwa, Ghana, MP, pp. 1–11.

Katsabanis, P.D. and Omidi, O., 2015. The effect of delay time on fragmentation distribution through small and medium scale testing and analysis. Fragblast 11. Proceedings of the 11th Int. Symposium on Rock Fragmentation by Blasting, AusIMM, Carlton, Australia, pp. 715–720.

Kulatilake, P.H.S.W., Hudaverdi, T. and Qiong, W., 2012. New prediction models for mean particle size in rock blast fragmentation. Geotech. Geol. Eng., pp. 1–23. https://doi.org/10.1007/s10706-012-9496-3.

Kuznetsov, V.M., 1973. The mean diameter of the fragments formed by blasting rock. Soviet Mining, 9, pp. 144–148.

Lu, W., Leng, Z., Chen, M., Yan, P. and Hu, Y., 2016. A modified model to calculate the size of the crushed zone around a blast-hole. Journal of the Southern African Institute of Mining and Metallurgy, 116, pp. 413–422.

Olofsson, S.O., 1988. Applied Explosives Technology for Construction and Mining. Nora Boktryckeri AB, Sweden, 315 pp.

Onederra, I., Esen, S. and Jankovic, A., 2004. Estimation of fines generated by blasting – Applications for the mining and quarrying industries. Mining Technology, 113(4), pp. 237–247.

Ouchterlony, F., 2003. Influence of blasting on the size distribution and properties of muckpile fragments, a state-of-the-art review. MinFo Project P2000–10: Energy Optimisation in Comminution, Lulea University of Technology, Sweden, 114 pp.

Ouchterlony, F., 2005. The Swebrec© function: Linking fragmentation by blasting and crushing, mining technology. Transactions of the Institutions of Mining and Metallurgy, Section A, 114, pp. A29–A44.

Ouchterlony, F. and Sanchidrián, J.A., 2019. A review of development of better prediction equations for blast fragmentation. Journal of Rock Mechanics and Geotechnical Engineering, 11, pp. 1094–1109.

Persson, P.A. and Holmberg, R., 1983. Rock dynamics. Proceedings of the 5th ISRM Congress, Melbourne, Australia, April, n. ISRM-5CONGRESS-1983-242.

Radziszewski, P., 2013. Energy recovery potential in comminution processes. Minerals Engineering, 46–47, pp. 83–88. Elsevier. www.elsevier.com/locate/mineng.

Rumpf, H., 1973. Physical aspects of comminution and new formulation of a law of comminution. Powder Technology, 7(3), pp. 145–159.

Rustan, A. and Vutukuri, V.S., 1983. The influence from specific charge, geometric scale and physical properties of homogeneous rock on fragmentation. Proceedings of the First International Symposium on Rock Fragmentation by Blasting, Lulea, Sweden, pp. 115–142.

Sanchidrián, J.A., 2010. Rock fragmentation by blasting. Proceedings of the 9th Int. Symposium on Rock Fragmentation by Blasting, CRC Press, Sanchidrián Ed, Granada, Spain, September 2009.

Savir, Z., Edri, I., Feldgun, V.R., Karinski, Y.S. and Yankelevsky, D.Z., 2009. Blast pressure distribution on interior walls due to a partially confined explosion. Proceedings of the International Workshop on Structure Response to Impact and Blast (IWSRIB), Haifa, Israel, November.

Scott, A., David, D., Alvarez, O. and Veloso, L., 1998. Managing fines generation in the blasting and crushing operations at Cerro Colorado Mine. Proceedings of the Mine to Mill 1998 Conference, The Australasian Institute of Mining and Metallurgy, Brisbane, Australia, pp. 141–148.

Shad, H.I.A., Sereshki, F., Ataei, M. and Karamoozian, M., 2018. Investigation of rock blast fragmentation based on specific explosive energy and in-situ block size. International Journal of Mining and Geoengineering, 52(1), pp. 1–6.

Siddiqui, F.I., Shah, S.M.A. and Behan, M.Y. 2009. Measurement of size distribution of blasted rock using digital image processing. Journal of King Abdulaziz University: Engineering Sciences, 20(2), pp. 81–93.

Silva, A.C.S., de Souza, J.C., Silva, R.A. and Rocha, S.S., 2016. Simulation and analysis of fragmentation in rock blasting at the Herval Quarry-Barrerios – Pe. 24th World Mining Congress, Rio de Janeiro, Brazil, October 18–21, pp. 231–241.

Stamboliadis, E.T., 2013. Energy distribution in comminution: A new approach to the laws of Rittinger, Bond and Kick. Canadian Metallurgical Quarterly-the Canadian Journal of Metallurgy and Materials Science, Taylor and Francis, pp. 249–258.

Szuladzinski, G., 1993. Response of rock medium to explosive borehole pressure. Proceedings of the Fourth Int. Symposium on Rock Fragmentation by Blasting-Fragblast-4, Vienna, Austria, pp. 17–23.

Vesilind, P.A., 1980. The Rosin-Rammler particle size distribution. Resource Recovery and Conservation, 5(3), pp. 275–277.

Chapter 4

Explosives and priming systems

4.1 Introduction

The supply of explosives is an important area for evaluation. There are many legal restrictions on explosives manufacturing and transportation worldwide. In many locations, the host country controls the movement of explosives through government regulations. In other countries, such as the United States, explosives are distributed by private corporations. However, regardless of the geographical area of reference, end-users must evaluate potential suppliers with regard to purchasing price, delivery modes, and evaluation of specific products marketed by the supplier to determine if the products will meet the needs of the mining operation and how the products will be transported to the job site (SME Handbook, 1996).

In the following, a synthetic classification of the explosive products, accompanied by a brief description, is provided.

The terminology used refers to the scientific literature, though no agreed uniformity among authors exists.

For detailed description of the individual explosives, it is suggested to examine the information made available by the manufacturers. In the following paragraph, the products are divided into two groups: explosives (the constituents of the charge) and priming devices (necessary to induce the explosion of the charge at the wanted time). Figure 4.1 provides the essential nomenclature of a blast hole.

4.2 Explosives

A simplified classification is shown below:

Deflagrating (black powder)
Detonating explosives

1) Nitro-glycerine-based explosives: dynamites, including granular dynamite (straight dynamite, high-density extra dynamite [ammonia dynamite], and low-density dynamite) and gelatine dynamite (straight gelatine dynamite, ammonia delatine dynamite, and semigelatin dynamite).
2) Water gels, emulsions, and slurries (wrapped or unpacked).
3) Dry blasting agents: poured or bulk ANFO, aluminized ANFO, densified ANFO, and packed (waterproof) ANFO.
4) Binary explosives: consisting of two-component products that are mixed on-site to obtain an explosive.

DOI: 10.1201/9781003241973-4

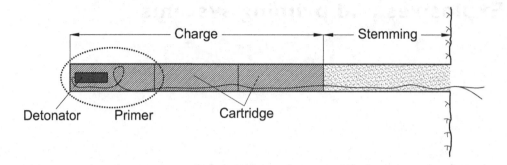

Figure 4.1 Essential nomenclature of a blast hole: hole; cartridges; "primer"; detonator.

Table 4.1 collects the most important characteristics of the main explosives used for industrial purposes.

According to the scheme above, the detonating explosives are listed in descending order of specific energy (MJ/kg), from top to bottom.

The most used oxidizer is ammonium nitrate NH_4NO_3, but in some compositions (conceptually falling within the extensive categories listed), sodium nitrate $NaNO_3$, potassium nitrate KNO_3, or even calcium nitrate $Ca(NO_3)_2$ are used for the same function.

Catalogs are available for all commercial explosives, containing the main data of interest. The nomenclature generally used for the different classes is evocative of the product (e.g., the suffix gel refers to gelatinized explosives), but, of course, it is necessary to check the technical characteristics to choose the most suitable product for the specific use-case. For packed products, cartridge sizes and weights are also available. An example, relating to ANFO explosives, is given in Tables 4.2 and 4.3.

Black powder

It belongs to the category of low explosives, which deflagrate rather than detonate. Its reaction rate ranges between 360 and less than 900 m/s. Black powder normally has little water resistance, is highly flammable, sensitive to an explosive cap of strength n. 6 (Tsakonas et al., 1979; Braithwaite et al., 2012), and has a lifting action due to blasting. Affiliated to low explosives, black powder generally does not induce rock fragmentation, but only a rough separation of the rock mass along the pre-existing discontinuities.

At present, black powder is applied for the detachment of blocks (regular or shapeless) in quarries of ornamental or construction stones. It is commonly used in a granular state (usually packed in cylindrical paper cartridges at the place of use) but can also be prepacked in cylinders. It cannot be used underground. It should be noted that although it is a deflagrating explosive, in many cases there is the need to trigger it by a detonation (detonating cord) to guarantee the simultaneous ignition of several charges. Anyway, the triggering performed by a detonation does not imply that the black powder will be transformed into a detonating explosive: the reaction obtained in terms of impact effect on the rock will be minimal, compared to detonating explosives.

Dynamites

This family of explosives is characterized by the highest specific energy (both by weight and volume), high detonation velocities, and high densities. When gelatinized (gelatinization is the process of stabilizing an explosive mixture/compound to make it safely transportable and

Table 4.1 Properties of the main industrial explosives.

Density (kg/m³)	Energy release MJ/kg	Detonation velocity (m/s)	Water resistance	Fume class
Straight dynamite				
1,400	3.9	5,200	Good	Poor
High-density ammonia dynamite				
1,300	3.9	2,450–3,800	Fair	Good
Low-density ammonia dynamite, high velocity				
1,200	3.7	3,350	Fair	Fair
1,000	3.7	3,050	Fair	Fair
900	3.6	2,900	Poor	Fair
800	3.6	2,600	Poor	Fair
Low-density ammonia dynamite, low velocity				
1,200	3.7	2,450	Fair	Fair
1,000	3.6	2,300	Poor	Fair
900	3.6	2,150	Poor	Fair
800	3.6	2,000	Poor	Fair
Blasting gelatine				
1,300	3.9	7,620	Excellent	Poor
Straight gelatine				
1,300	3.9	7,010	Excellent	Poor
1,400	4.1	6,100	Excellent	Good
1,500	4.7	5,030	Excellent	Good
1,700	4.5	3,350	Excellent	Good
Ammonia gelatine				
1,300	4,3	6,100	Very good	Good
1,400	4,1	5,340	Very good	Very good
1,500	4,7	4,900	Very good	Very good
Semigelatin				
1,300	3,7	3,700	Very good	Very good
1,200	3,7	3,700	Very good	Very good
1,100	3,4	3,500	Good	Very good
900	3,2	3,200	Fair	Very good
Emulsions				
1,200 ± 50	3,0	4,900–5,400	Very good	Very good
1,150 ± 50	3,8	4,770–5,200	Very good	Very good
1,150 ± 50	4,1	4,500–5,000	Very good	Very good
1,250 ± 50	4,4	6,200–6,300	Very good	Very good
Water gels and slurries				
1,050 ± 50	2,7	4,500–4,900	Poor	Poor
1,050 ± 50	2,7	4,550–4,900	Poor	Fair
1,000–1,200	3.12	4,500–5,000	Fair	Fair
1,000–1,200	3.58	3,700–4,500	Fair	Fair
1,000–1,200	3.52	3,900–4,600	Fair	Fair
1,000–1,200	3.44	4,100–5,000	Fair	Fair
1,000–1,200	3.39	4,300–5,300	Fair	Fair
1,000 ± 50	2.7	3,950–4,050	Poor	Poor
1,250	3.5	5,000	Good	Fair
1,300	4.3	5,500	Good	Fair
ANFO				
800 ± 50	3.9	2,200–3,050	Poor	Good
850	3.9	4,000	Poor	Good
Binary explosives				
820 ± 50	4.3	>4,000	Fair	Good
1,150	3.9	4,300	Good	Very good
1,150 ± 50	3.5	4,770–5,200	Very good	Very good
1,150 ± 50	3.4	4,500–5,000	Very good	Very good
1,250 ± 50	3.5	6,200–6,300	Very good	Very good

Table 4.2 Technical characteristics of an ANFO-type explosive

Technical characteristics (nominal values)	
Density	0.8 g/cm3
Velocity of detonation	4,000 m/s
Energy release	3.9 MJ/kg
(REE-WS) (ANFO 100%)	100%
(REE-BS) (ANFO 100%)	100%
Gases volume	978 l/kg
Fumes	Between 2.27 and 4.67 l/100 g

Source: www.maxam.net/en/maxam/contacta_maxam

Table 4.3 Packing and weight packing of an ANFO-type explosive.

Formats and crates (nominal values)				
Packing		Weight packing (kg)		
Plastic bags		25 kg		
diameter length (mm)	weight/cartridge (g)	n° cartridge box	weight/box (kg)	type of cartridge
50 × 490	833	30	25	Plastic film
55 × 490	962	26	25	Plastic film
65 × 490	1,250	20	25	Plastic film
75 × 490	1,563	16	25	Plastic film
85 × 490	2,083	12	25	Plastic film
125 × 490	5,000	5	25	Plastic film

manageable), dynamite has excellent water resistance and ease of priming. Concerning the economic aspect, their cost is quite high, making their usage infrequent. However, often dynamites are used to provide only part of the total charge (Cook et al., 1962; Agioutantis et al., 2001; Wang et al., 2003).

Working with dynamites requires special attention to conservation problems (temperature control of storage rooms, ventilation; see Cavagna et al., 1967), since they exhibit a greater sensitivity to impact compared to other explosives (Nabiullah and Singh, 1990; Hu et al., 2001). Furthermore, this explosive should be handled with cautious gloves, as nitro-glycerine is absorbed by the skin and can pass into the blood causing disturbances (even very intense headaches, fortunately not permanent).

A schematic classification of dynamites is provided in the following:

1) Blasting gelatine: It is one of the strongest commercial products, made up of 92–94% of NG gelatinized and 6–8% of nitrocellulose. Being the most energetic product available on the market, it is mainly used in the seismic industry, although some commercial products are also used as primers and boosters.
2) Straight dynamite: Its energy source comes from NG, SN, and AN, including absorbents such as wood pulp and flour that also act as combustibles. These dynamites are available on the market in different weight strengths (40, 50, and 60% are the most common) which

correspond to the weight percentage of NG contained in the formulation. Ditching dynamite (50% NG) is a commonly used product. These dynamites are characterized by high reaction velocity and brisance, low flame temperature, and good water resistance; however, they are sensitive to shock and produce poor fume quality.

3) Ammonia dynamite (or "extra" dynamite): It is a granular mix that contains a smaller quantity of NG mixed with AN and SN. Two varieties are available on the market which are functions of the density. Considering the high-density ones, these explosives are characterized by a high-velocity product (detonation velocity of 2,700–6,000 m/s), fair to good fumes, and low explosion temperature. Products available on the market have weight strengths of 20–60%. They have poor water resistance.

Considering the low-density variety, these explosives come in either a high- or low-velocity series and a constant 65% weight strength. These products have detonation velocities ranging between 2,200 and 4000 m/s, fair to good fumes, and poor water resistance. It should be reported that there is another category of explosives belonging to the low-density variety: the permissible type. These products are like the low-density ammonia dynamites except that they contain cooling salts, such as sodium chloride. Before usage, permissible explosives must be approved by the U.S. Bureau of Mines under specified usage conditions (Siskind and Kopp, 1995; Santis et al., 1995; Mainiero and Verakis, 2010). This material usually has good fumes and fair to poor water resistance.

4) Gelatin dynamite: It contains nitro-glycerine (from 60% up to 92%) gelled with nitrocellulose and various absorbent filler materials that provide the final product with a rubber-like consistency that is water-resistant.

Gelatin dynamites can be classified as straight gelatin or ammonium (extra) gelatin. The base composition for both kinds is the same as the straight ammonia (extra) dynamites with the addition of nitrocellulose for obtaining a gel consistency. Two different gelatin dynamites are available on the market: straight gelatin and ammonium (or Special) gelatins. Straight Gelatin has a detonation velocity of 4,000–7,000 m/s. Varieties with a strength rating above 60% have poor fume characteristics. Water resistance is excellent, and the material is very compact. As concerning ammonium gelatins, their composition is like straight gelatin except for the replacement of a certain amount of nitro-glycerine with ammonium and sodium nitrates and carbonaceous fuels. It has a detonation velocity between 3,000 and 7,000 m/s. Water resistance is good. Ammonia gelatins are commonly used as boosters and primers to initiate cap-insensitive explosives.

5) Semi-gelatin dynamite: It is a mixture of ammonia gelatin and ammonia dynamite, with lower energy than gelatin dynamites, but it preserves good water resistance. The detonation velocities are between 3,000 and 4,500 m/s. The fume rating is good.

Semi-gelatins are ammonia gelatin with a small amount of nitrocellulose and a 65% weight strength. They are also used as primers and boosters.

Blasting agents

A blasting agent is any material or mixture consisting of a fuel and oxidizer that is intended for blasting and not classified as an explosive. A blasting agent consists primarily of inorganic nitrates (ammonium and sodium nitrates) and carbonaceous fuels. The addition of an explosive ingredient (i.e., TNT) in a sufficient quantity changes the classification of the mixture from a blasting agent to an explosive. If a blasting agent is initiated under unconfined conditions, it cannot be detonated by means of a No. 8 test blasting cap (Watson et al., 1975; Meyers et al., 2001; Kim et al., 2021) unless an explosive ingredient or sensitizer is added. It should be reported that No. 8 test caps contain the equivalent of 2 g of a mixture composed of 80% of mercury fulminate and 20% of potassium chlorate. Blasting agents may be classified as (1) dry blasting agents, (2) emulsions, (3) water gel, or (4) slurry blasting agents. Products such as those mentioned below are also common:

Bulk mixed compounds: This group of blasting agents includes most of the ammonium nitrate-fuel oil mixtures and bulk slurries, which are often mixed on the job site by the supplier in the delivery truck. The detonation velocities range between 2,700 and 4,600 m/s. These products have no water resistance with fair to good fumes if properly mixed and detonated.

Pre-mixed nitro-carbo-nitrates (NCN): These blasting agents include products prepared by a commercial manufacturer and purchased by consumers in the ready-to-use package form. The densities of these products are variable. Velocity ranges from 3,600 to 4,600 m/s.

Cycling of ammonium nitrate: Ammonium nitrate releases more gas upon detonation than any other explosive due to its density. In its pure form, ammonium nitrate is almost inert (inactive). Its chemical composition with reference to the unit of weight is approximately 60% oxygen, 33% nitrogen, and 7% hydrogen. Two characteristics make this compound unpredictable and dangerous (Sjolin, 1972). The former characteristic relates to ammonium nitrate and its water solubility: if uncoated, it can attract water from the atmosphere and slowly dissolve itself (Chaturvedi and Dave, 2013). In order to prevent or at least minimize this phenomenon, prills are produced with a protective coating of wax or clay that acts as a moisture retardant. The latter and most important characteristic reflects a phenomenon called "cycling": the ability of a material to change its crystal form with temperature. Ammonium nitrate will have one of five crystal forms depending on the temperature, but the cycling variation can seriously affect both the storage and performance of any explosive containing ammonium nitrate (Ingman et al., 1982; Dellien, 1982).

Considering the occurrence of the cycling phenomenon, two different scenarios can occur in function of the waterproofing level of the ammonium nitrate. The effect of cycling of ammonium nitrate when isolated from the humidity in the air is that the prills break down into finer particles or enlarge till the point at which they are nearly inert: as it can easily be deduced, the density changes. More into the details, when the temperature exceeds 32°C, the prills break down into smaller crystals and cause a density increase from 0.8 to 1.2 g/cm3. As a result of the density increase, the compound's detonation velocity will increase too: from 3,000 m/s for a density close to 0.8 g/cm3 to 4,500 m/s, typical detonation velocity for a density close to 1.2 g/cm3. Under the previous condition, there is the risk of attaining the critical density, and beyond this limit, the explosive will no longer detonate (Nie et al., 1993; Haskins and Cook, 2007). Analyzing the second condition, namely when some blasting agents made up of ammonium nitrate are not sealed enough to exclude humidity, a different scenario occurs: after the ammonium nitrate has undergone one cycle, the waterproof protective coating is broken, and the water vapor in the air condenses on the particles. As cycling continues and more water accumulates, the mass of ammonium nitrate starts to dissolve, during which it starts to recrystallize into large crystals. For instance, considering ANFO, after cycling, it may no longer be homogenous: very dense areas can be present close to areas of large crystals (hence characterized by a lower density). The performance of this product may range from that of a very powerful explosive to one that just burns or one that will not detonate at all (Johnson et al., 1983; Salyer et al., 2010; López et al., 2013).

Most dynamites, both regular nitro-glycerine and permissible, contain some percentages of ammonium nitrate. Blasting agents, instead, are almost totally composed of this compound (Dick, 1972; Juchem et al., 2015). For this reason, cycling does not affect dynamite strongly as in the case of ANFO. The two temperatures at which cycling will occur under normal conditions are −18° and 32°C. Consequently, it can be stated that products stored over the winter or during the summer are most likely to undergo some amount of cycling. Furthermore, it is noteworthy to mention that if the storage area is poorly ventilated during the summer, the cycling temperature may be reached daily.

ANFO

The acronym (ammonium nitrate-fuel oil) describes the explosive mixtures that have gained popularity in open-pit excavation operations since the 1950s (to a lesser extent in underground work).

Ammonium nitrate is probably the cheapest oxidizing salt, and since the 20th century, it has been used (mainly to correct the oxygen balance) as an ingredient in industrial explosives. In the 1950s, simple mixtures of ammonium nitrate (fertilizer) with solid fuel (pulverized coal), known as Akremites, were experimented and had some diffusion (Cardu, 2007); the transition to mixtures of simpler nitrate-liquid fuel preparation (the ordinary gas oil), designated by the initials ANFO, was almost immediate, initially prepared with crystalline agricultural nitrate, then with nitrate expressly produced in the form of "prill" (porosity grains and controlled particle sizes).

ANFOs can be added with aluminum powder (particularly effective fuel), naturally in a balanced amount, to increase the explosion temperature and, therefore, the pressure obtainable. These types of ANFO are called metalized ANFOs (Dick et al., 1983; Oommen and Jain, 1999; Balachandar and Thangamani, 2019; Elbeblawi et al., 2022).

On the contrary, when there is the need to reduce the explosion pressure (e.g., in the case where a release effect with minimal fragmentation is desired), porous fuel can be added to ANFO, obtaining the reduction of the density and therefore the explosion pressure. This type of ANFO is called lightened ANFO (styranol). The expanded polystyrene granules are a typical additive used for lightening the explosive.

ANFOs are explosives of low cost, low density, specific volumetric energy lower than that of the explosives described previously. The main weakness of ANFOs is their inability to resist water action, indeed, their water tolerance is null. Thus, with a density lower than 1 kg/L, holes containing water cannot even be loaded. ANFOs can hardly detonate in holes smaller than 50 mm. For a proper and complete detonation (also in holes of bigger diameter), it is always suggested to use a booster, that is, a more sensitive auxiliary explosive charge (e.g., dynamite).

ANFO is mainly used in open-pit operations, but they are also used underground (e.g., in Norway they are widely used in the excavation of tunnels, even in tough rocks). The problem with using ANFOs is the presence of water, though it can be neutralized with properly designed mixed loads.

In Italy, only ANFOs prepared in authorized factories and packed in cartridges are allowed. However, pneumatic loading of the bulk mixture is authorized if the used machine is certified. In other countries, on-site preparation is allowed: the mixing of ammonium nitrate and fuel oil is carried out by using a piece of approved equipment and the bulk loading can be carried out also by gravity

(Watts, 2001; Sapko et al., 2002; Fleetwood et al., 2012; Silva and Orlandi, 2013). In this last case, it should be reported that the optimum value of the load density ranges between 0.8 and 0.9 kg/l.

Explosives with nitrates in solution (water gels, emulsions, and slurries)

They spread shortly after the success of ANFO, starting from the 1960s, initially only for bulk loading in large blast holes, and later also for loading ordinary blast holes in cartridges (Borg, 1995; Taguchi et al., 2005).

The development of these explosives was initially driven by the need to overcome the two weaknesses of the ANFO, that is, the low density and the strong sensibility to the water, keeping the substantial cost-effectiveness of the formulation.

The first types of slurries were simply made up of suspensions of ammonium nitrate and explosives (usually TNT) in a saturated solution of ammonium nitrate: the explosive acted as both a sensitizer and a fuel.

Subsequently, other sensitization systems were developed (dispersion of air bubbles, hollow glass microspheres, etc.) and stabilization systems (gelatinization with the addition of organic gels) so products that do not contain explosive substances were obtained. These products contained only oxidizers and fuels and were similar in terms of features to dynamites but with considerably lower specific energy than the latter. They are often appointed as water gels (Kane, 1979; Mohan and Srinivas, 1980; Persson, 1980; Persson et al., 1980; Wada et al., 1991).

Lastly, the system was further improved thanks to emulsification (explosive emulsions) and mixing the emulsions with ANFO, obtaining the heavy ANFO (Coleman and Larson, 1994; Hudaverdi et al., 2009; Leidig et al., 2010; Shao and Feng, 2015; Žganec et al., 2016).

Therefore, there is at present times a wide range of explosives that are very safe to handle, with a good water tightness, characterized by slightly low specific energy, but still suitable for all uses (both in open pits and underground) at low costs.

The triggering is more difficult than "classic" explosives (such as dynamites); consequently, it is advisable to insert a booster of dynamite cartridge equipped with a detonator in the charge. However, it should be reported that for certain types of explosives, the use of the detonator alone guarantees the hole blasting. According to the common explosive nomenclature, the couple composed of a booster and a detonator is called primer. The critical diameter is higher than that of dynamites. In some types, desensitization can occur due to excessive compaction.

In Italy, these explosives can be only used in cartridges. Out of the country, loading in bulk, by casting or pumping, as well as on-site mixing of components, are allowed and commonly adopted (also with the possibility of changing the composition of the charge along the hole). Of course, these practices are convenient only in the case of very high consumption, which requires ad hoc equipment installed on special vehicles.

High-safety explosives

The compositions can considerably vary from type to type, but synthetically they could be defined as "diluted dynamites".

These explosives can be used in construction sites where the development of explosive reactions is feared; consequently, it can be stated that their main market is represented by coal mines.

Dilution with cooling salts necessarily implies a decrease in specific energy and pressure from the flames' cooling (Ringgenberg and Mathieu, 2004; Nikolczuk et al., 2019). The need for

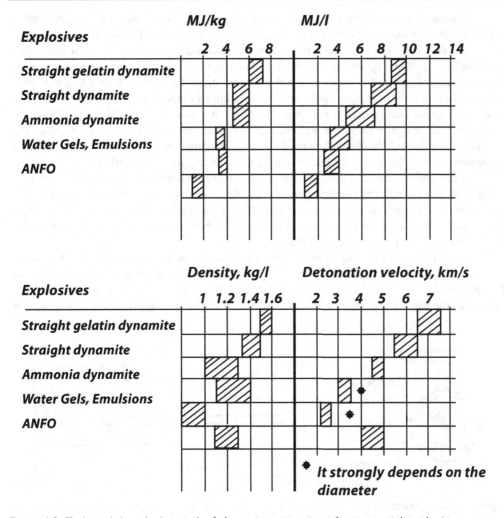

Figure 4.2 Technical data (indicative) of the main categories of commercial explosives.

good detonation on small diameters and a positive oxygen balance makes the presence of nitroglycerine almost inevitable. However, there are also types that do not contain this component (Kholodenko et al., 2014).

The ranges of deviation of the most interesting characteristics pertaining to the products examined up to now are listed in Figure 4.2.

4.3   Fuses and detonic tubes

In blasting works, it is necessary to manage the timing of in-the-hole charges that communicate with the outside through a narrow opening (the hole itself). To solve this matter, very thin charges ("cores") covered and secured by flexible casings known as "fuses" were developed.

Despite being similar in appearance, two types of fuses are distinguished: the ordinary fuses and the detonating cords.

The ordinary fuses are also called safety fuses. Their core is of phlegmatized deflagrating explosive (black powder), and their reaction speed is very low (about 1 cm/s). The casing of the safety fuse is waterproofed, so it also burns underwater (but does not have, of course, unlimited water tightness). When ignited at one end, the slow reaction passes through the fuse without a visible external flame and re-emerges as a flame "dart" from the other end, to induce the defla-gration of a charge or to trigger a primary explosive linked to the fuse (Figure 4.3). The safety fuse is an outdated device and is not recommended, but it can still be applied at small scales, when the blast design requires the triggering of only one detonator, or, more often, when the blasting circuit is made by detonating cord: in fact, detonating cord can transfer the detonation wave generated by the initiated device to any distance as required (Zou, 2017).

As for detonating cords, the core consists of a detonating explosive (Pentrite – PETN), with a detonation speed of 6–7 km/s (see Figure 4.4). The detonation occurs by triggering a detonating charge in contact with one of its ends and exploding it triggers the detonation of other charges with which it is in contact. Differently from the safety fuse, the detonation of the detonating cords causes a clearly visible external effect.

The linear charges of the most used types are between 10 and 20 g/m of PETN. Detonating cord is widely used in open-cast works, rarely underground. In addition to being a device to transmit the detonation to the charges of many blast holes, or to ensure the complete detonation of a blast hole when it is charged according to decks, it can be used into the holes alone: it is, in

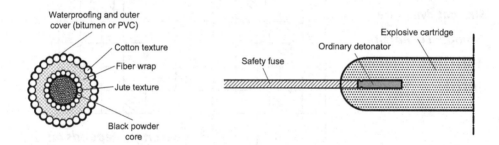

Figure 4.3 Left: structure of the safety fuse; right: example of initiation of an ordinary detona-tor by means of a strand of safety fuse.

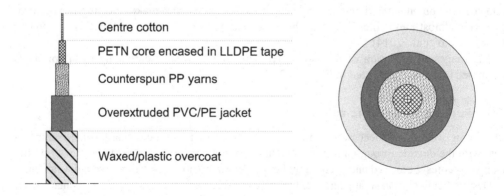

Figure 4.4 Structure of the detonating cord

fact, made up by an elongated charge of small diameter, high detonation velocity, and it is highly suitable for blasting works where only the split effect is desired, without rock comminution (see an example in Figure 4.5).

Detonic tubes, Nonel tubes, and shock tubes cannot be considered fuses, regardless of the similarity of their function. They consist of flexible tubes containing a small amount of detonating charge (about 20 mg/m), which remains confined in the tube itself upon detonation. Detonating reaction travels with a speed close to 2 km/s without causing any visible external effect (see Figure 4.6). The detonation is triggered, similar to the detonating cord, by a starter unit in contact with the tube, but it remains confined inside the tube until the impulse is transmitted to

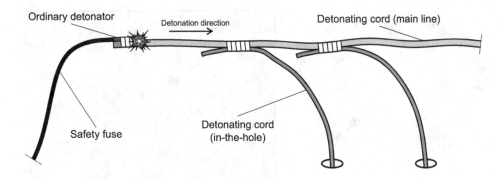

Figure 4.5 Example of an instantaneous blast with detonating cord. The initiation is provided thanks to an ordinary detonator, triggered by a strand of safety fuse.

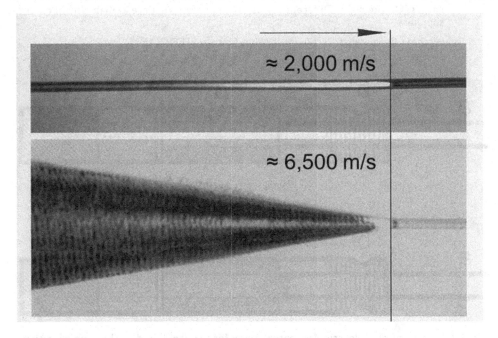

Figure 4.6 Shock tube (on the top) versus detonating cord (on the bottom)

Source: Product Manual Nonel and Primacord (2011).

the detonator at its other end (differently, the detonating cords triggers all charges in which it is in contact).

The acronym Nonel means non-electric. Nonel tubes are marketed in various lengths, already equipped with detonators and surface connectors. Circuits suitable for triggering the wanted number of blast holes according to the established sequence can be easily designed and realized.

In summary, Nonel tubes are to be considered non-alterable appendices of their detonators, and the Nonel blasting system is not "built" but rather "assembled" on-site with precast elements (an example is provided in Figure 4.7). A sketch of a non-electric detonator is provided in Figure 4.8.

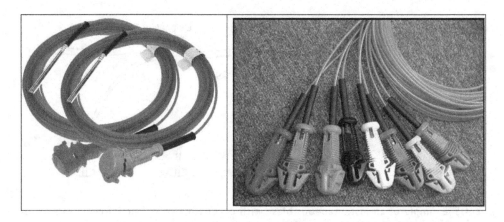

Figure 4.7 Non-electric initiation. On the left: a combination of surface connectors and detonators. On the right: Different styles of connectors that are available from different suppliers

Source: Pan-European Competence Certificate for Shot-firers by EFEE (2019)

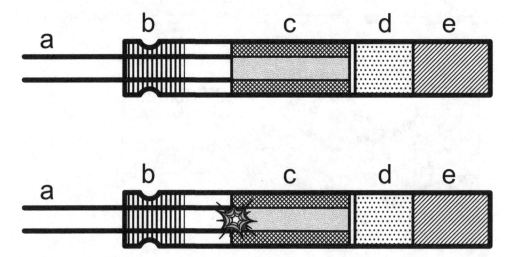

Figure 4.8 Section of non-electric detonator, ready for use (top), and at the time of initiation (bottom); a: shock tube; b: sealing plug; c: delay element; d: primary charge; e: base charge.

The Nonel system is widely spread around the world. It is used in both civil excavations and in quarries and/or mines, both underground and open pits.

It has the great advantage of allowing the triggering of a very high number of charges according to an accurate sequence (see the example in Figure 4.9).

Two lines of history

Black powder has been used since ancient times in China for rocket launches and since the late 1200s in Europe to load weapons. These two uses do not require ignition from a distance, as the operator is protected by the rocket wall's thickness or the weapon's barrel, which "should not" explode.

The case of blast holes is different, although they have been used at least since 1400 for military and civil purposes: in this case, the operator must be far from the charge that requires initiation. The charge used to be connected to a sheltered spot where the operator ignites it with straws full of black powder, through which the flame reaches the charge. This rudimentary, uncertain, and dangerous system was used until the 19th century. Only in 1831, Bickford did invent and produce the safety fuse, with a waterproof and flexible case and controlled burning rate, which almost immediately replaced open flame systems (Meyers and Shanley, 1990; Fodor, 2010).

In the second half of the 19th century, solid detonating explosives began to appear on the market. During the same period, thin and protected charges of detonating explosives (detonating cords) were also being produced: the aim was to allow the transmission of the detonation, also at certain distances, with minimum delay from one charge to another. Detonating cords are still widely used, yet an evolution has occurred in this field that led to the transition from the "black powder furrow" to the safety fuse. The common detonating cords are free detonation devices, that is, with an external effect, but ever since the mid-1900s, systems have also been available in which the detonation can be confined in a tube. In this last version, the combustion of the powder in the fuse remains confined in an envelope. The detonic tubes allow a progressive approach to the ideal one, referred to in the sketch of Figure 4.10.

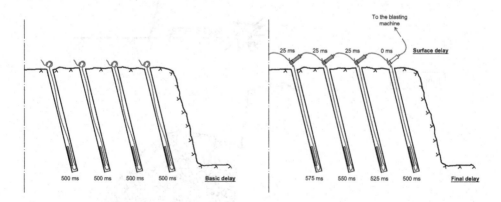

Figure 4.9 Principle of initiation of a blast using Nonel detonators. Left: Long-delayed detonators (500 ms) are placed at the bottom of the holes. Right: connection of Nonel tubes with short delay (0 and 25 ms) surface connection units; in this way, the first blast-hole detonates after 500 ms from the triggering, the others with a progressive timing of 25 ms. Basically, the blast is micro-delayed with a delay scale of 25 ms, and the first blast hole detonates only after all surface connection units have been activated, without any risk of interruption/breakage of the circuit.

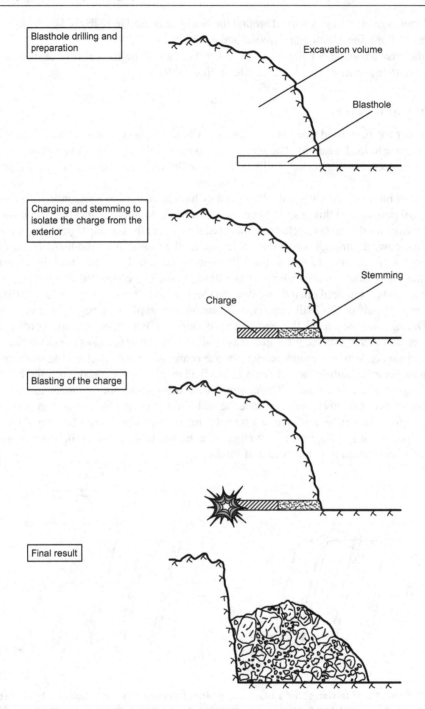

Figure 4.10 Ideal solution to the blasting problem.

4.4   Detonators – general

The detonators or "caps" have been developed to control explosive energy release in a productive and safe sequence. They time the sequence of detonation of multiple charges in a blast to control the blast's breakage, rock movement, and adverse environmental effects according to the desired purposes. The various systems transmit a signal from hole to hole or charge-to-charge with varying timing accuracy in short intervals. These intervals can vary in duration from milliseconds to seconds (ISEE Blasters' Handbook, 2011). Each initiation system has unique physical properties, performance characteristics, and applications (ESSEEM Project, 2013).

The detonation is usually produced by a small charge of the primary explosive, even though the so-called detonators without primary (NP) have been developed and are also in use. NP detonators are made up of a charge of secondary explosive (PETN) with suitable grain size, suitable density, stratification, and confinement: working in this condition, the reaction of deflagration almost instantaneously moves to a state of detonation, behaving as a primary charge. This type of detonator was developed to achieve a lower sensitivity to accidental impacts than the normal detonators containing primary explosives.

Other types of detonators also exist, such as those containing only a secondary charge called EBW. They are mainly used for research or military purposes, and the initial impulse is given by a real "electric detonation", that is, the explosion of a metal wire induced by a very high-intensity current pulse (Varesh, 1996; Lee et al., 2014a, 2014b, 2014c; Rae and Dickson, 2019).

Returning to detonators with a primary charge, they were originally made up of mercury fulminate, often called "fulminate" in technical jargon, although this substance is no longer in use, and has been replaced by others (lead azide, lead styphnate). Furthermore, the original detonators (such as those of Nobel) contained only primary explosives, but the current ones contain a two-layer charge: one of a primary, which receives the initial stimulus, and the other of a secondary (the so-called base charge, generally PETN) which explodes to detonate the first.

An ordinary detonator contains in total a bit less than 2 g of explosives (between primary and secondary charge). Conventionally, for storage regulation purposes, it complies with 2 g of category II explosives, according to the Consolidated Act of National Public Security Laws – TULPS (Dadone, 2004; Rossetti et al., 2017).

The structure of an ordinary detonator is portrayed in Figure 4.11.

4.4.1   Detonators: initiation

In the described ordinary detonator, the detonation is caused by the "dart" (jet of flame) of the fuse, which strikes the primary charge (Figure 4.12).

Ignition with a safety fuse is inconvenient and dangerous, especially if more detonators (i.e., more blast holes) must explode (simultaneously or gradually) and is commonly replaced by other systems, such as electric, non-electric, or electronic detonators.

4.4.2   Electric detonators

The detonator, upstream of the primary charge, is equipped with a special head, consisting of a small deflagrating charge, including an electric resistance. The resistance is heated by the

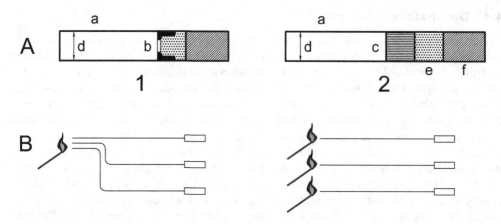

Figure 4.11 Scheme of timing systems with an ordinary detonator and safety fuse.

    A. Types of ordinary detonators. 1. With safety fuse locking ring. 2. Without safety fuse locking ring. a: cap (Al, Cu, bronze, brass); b: fuse locking ring; c: deflagrating charge; d. internal diameter, corresponding to the external diameter of the fuse; e: primary explosive (mercury fulminate, lead azide, lead styphnate, etc.); f: secondary detonating explosive.

    B. Timing systems (blasting of multiple charges at pre-set intervals) achievable with the ignition of the safety fuse: on the left, simultaneous initiation of several fuses of different lengths; on the right, initiation of several fuses of equal length in succession. The numbers show the order of the explosion.

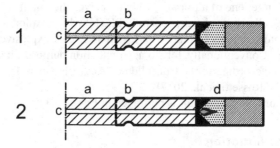

Figure 4.12 Ignition of an ordinary detonator. 1. Section of the device ready for use. 2. Section of the device at the instant of ignition. a: safety fuse; b: locking the detonator on the fuse, with special pliers; c: core of the fuse; d: dart that hits the primary charge, causing the detonation.

passage of a suitable electric current (ignition current), causing the deflagration of the integrated charge. This deflagration causes the primary charge to detonate (Figure 4.13).

    The detonator, shown in the scheme of Figure 4.13, is an instantaneous electric detonator. The term "instantaneous" refers to the fact that the explosion occurs with a negligible delay (about one millisecond, due to the thermal inertia of the wire and the reaction time of the explosive charge) with respect to the transmission of current. Electric detonators are produced with different "sensitivities", that is, they require different current intensities for their safe ignition.

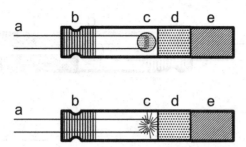

Figure 4.13 Section of an instantaneous electric detonator, ready for use (top), and at the time of initiation (bottom). a: leg wires; b: insulating cap; c: head (resistance, or strand, embedded in a deflagrating substance); d: primary charge; e: base charge.

Electric systems require an electrical power source to initiate detonators. Electric detonators are connected in circuits to provide them with an electric current. Sensitivity and resistance are not really characteristics of the detonator, but only of its ignition system. In fact, they do not influence the triggering power.

The definitions of high, medium, and low sensitivity (which correspond respectively to low, medium, and high current intensity) can reflect different ignition amperage operative values in relation to the different conventions followed by manufacturers. In rough, there are high-sensitivity (or low-intensity) detonators when the recommended current for safe ignition is around 1 A (but there is already a certain risk of ignition with extended passage of lower currents, around 0.2 A), medium sensitivity (or medium intensity) when the recommended current is around 2–3 A (and there is no risk of ignition at about 0.5 A), and low sensitivity (or high intensity) when the recommended current is of tens of A (20–30) and the possibility of ignition is limited to currents of a few A (3–4).

The electrical resistance of the system (obviously, one of the connection wires depends on their length, which varies according to the geometry of the circuit) varies directly with the sensitivity of used detonators. In principle, more sensitive detonators have greater resistance: high sensitivity types have a resistance of 1–2 Ω, those with medium sensitivity a few tens of Ω, and those with low sensitivity, a few hundreds of Ω.

Sensitivity and resistance are very important information that must be known by users. Both sensitivity and resistance are controlled and guaranteed by the manufacturers. The knowledge and the subsequent choice of sensitivity are crucial for deciding whether to use a given type of detonator where there is a fear that the circuit is affected by extraneous currents. As for the resistance, the correct computation of the whole circuit's electrical resistance is fundamental for verifying the correctness of the circuit and the suitability of the electrical source selected for priming.

Delayed detonators contain, above the primary charge, a pyrotechnic delay element that burns at a known rate and whose length and composition control the transit time of the burning front. Detonators classified as instantaneous, or "zero" delay don't contain a delay feature. In most cases, however, it is not necessary to detonate a single charge, nor to detonate several charges at the same time, but to detonate numerous charges according to a predetermined sequence, which means timing the explosions.

In most cases, therefore, delay scales are used, which vary according to the manufacturers and/or the country in which the detonators are marketed. Thus, for a blast to occur in sequence, detonators with different delay times must be used in the circuit (Figure 4.14).

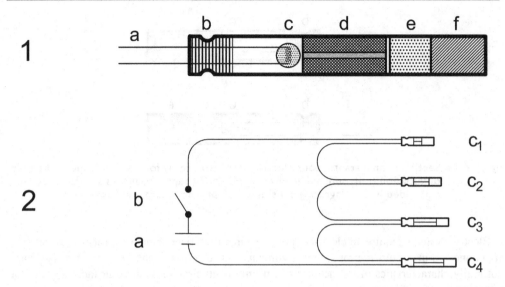

Figure 4.14 Schemes of a delayed electric detonator (1) and example of a timing system in electrical initiation (2).

1. Section of a pyrotechnic delayed electric detonator; a: leg wires; b: crimps; c: bridge-wire in fuse-head; d: pyrotechnic delay element (tube containing a deflagrating substance at a known and controlled reaction rate); e: primary charge; f: base charge.
2. Timing of explosions with delayed electric detonators; a: exploder; b: switch; c_1, c_2, c_3, and c_4: electric detonators with different delays (in the example, they are connected in series).

A series of detonators of this type can be used with the timing of 500 or 250 ms (ordinary delay series). Shorter delay scales are also available, equal to 100, 50, and 25 ms. Commonly used detonators have a delay sequence between 20 and 30 ms, depending on the different manufacturers.

4.4.2.1　Basic circuitry

Detonators are connected to power circuits at the appropriate time by portable generators or exploders. The circuits consist of a "disposable" part, that is, the leg wires supplied with the detonators, and a reusable part, which safely connects the blasting area with the site from where the operator transmits the electrical triggering pulse. Three basic electric circuits are used: series, parallel, and series-in-parallel circuits. The series circuit is the simplest since the current has only one flow path to follow (Figure 4.15). The parallel circuit is slightly more complex, being characterized by multiple flow paths for the current (Figure 4.16). It was used in the past to facilitate wiring in many high-speed tunnel and shaft sinking operations (ISEE Blasters' Handbook, 2011). The series-in-parallel circuit (Figure 4.17) is the most complex electric scheme since it combines the series and parallel circuits by replacing the single detonators with a series circuit; its main advantage is that many detonators can be fired from a blasting machine without a large voltage requirement.

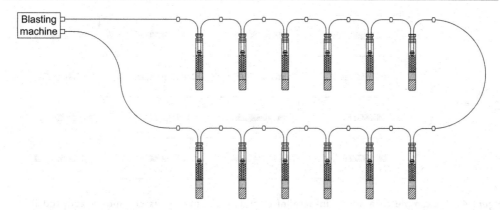

Figure 4.15 Example of a series circuit. The resistance of the circuit is calculated as the sum of the resistances of each detonator. $R_{tot} = \sum_{i=1}^{n} R_i$ where R_{tot} is the total resistance of the circuit due to detonators and n is the number of detonators. The disadvantage of this circuit is that, especially if low-intensity detonators are used, the blast is composed of many blast-holes, and the total resistance is very high.

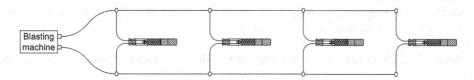

Figure 4.16 Example of parallel circuit; it is more laborious than the previous one in terms of organization but generates a lower circuit resistance: it is calculated as the ratio of a detonator's resistance to the number of detonators. $\dfrac{1}{R_{tot}} = \sum_{i=1}^{n} \dfrac{1}{R_i}$ where R_{tot} is the total resistance of the circuit due to detonators and n is the number of detonators.

The design and check of the circuits are based on elementary notions of electrical engineering; however, it is necessary to respect some important conditions:

- All the detonators of the same circuit must have the same electrical characteristics and, in mixed connections, all the groups of detonators in series to be placed in parallel must have the same resistance, which means that the overall circuit must be "balanced".
- It is not allowed to use, as a part of a firing circuit, an electric line that is also used for another purpose.

Both exploders and control devices must be approved: in particular, using an ordinary tester to check a firing circuit is extremely dangerous.

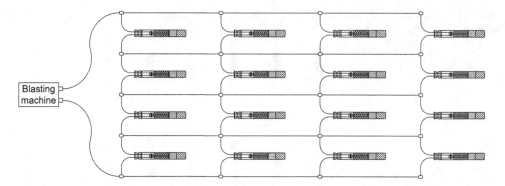

Figure 4.17 Example of a series-in-parallel circuit; this solution is commonly adopted if the blast consists of many blast holes. In these cases, the scheme must be preliminarily simplified according to the rules of electrical engineering, which respects a total resistance of the circuit equal to $R_{tot} = \dfrac{R_S}{n_S}$.

4.4.3 Electronic detonators

In the early 1990s, electronic detonators began to be industrially produced. These detonators are characterized by timing that is achieved in a completely different way. The detonator contains (in addition to the charge and the ignition head)

- a device (capacitor) designed to accumulate the electrical energy necessary to obtain the ignition;
- an electronic timer that allows energy discharge on the head at a programed time, with considerable precision (± 1 ms).

Once the holes are charged, the detonator is "loaded" and "programed" to detonate at the desired time. These devices are yet not widespread in some countries (such as Italy) due to their high cost. In short, with regard to core design and functionality, the delay is not pyrotechnic, but it is achieved electronically (Figure 4.18). There are more than ten different designs available worldwide, but the structure is basically the same: a computer chip controls delay times, which uses electrical energy stored in one or more capacitors to provide power for the timing clock and initiation energy.

The advantages of electronic detonators, in addition to greater safety, include accurate excavation profiles with a reduced back-break, better fragmentation reduction of fines, better control of vibrations, the possibility of obtaining an unlimited number of delays, a reduction in the ratio of drilled meters/costs. On the other hand, some disadvantages still have not been completely overcome, such as high costs and complexity of use.

In some cases, it makes sense to combine different kinds of initiation systems, for example:

- Accuracy: in Tunneling, the high precision of electronic detonators helps avoid over-break in the perimeter; the major (center) part of the whole circuit contains non-electric detonators.

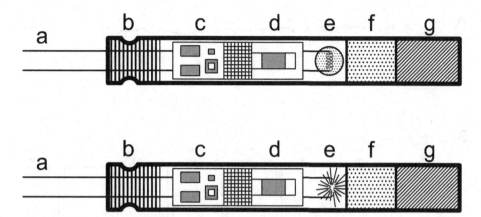

Figure 4.18 Section of an electronic detonator, ready for use (top), and at the time of initiation (bottom); a: leg wires; b: insulating cap; c: delay module (chipset) d: capacitor; e: head (resistance, or strand, embedded in a deflagrating substance); f: primary charge; g: base charge.

- Electricity: stray current (or the probability of lightning) forces to use non-electric detonators – for testing prior to blasting, the non-electric round will be initiated through an electric firing circuit.
- Testability: in demolitions, the two-way communication of electronic initiation allows testing the most important part of the whole circuit.

4.4.4   Nonel detonators

The Nonel system is probably the one among the affordable systems that allows the greatest versatility in timing. In fact, in this case, at least two different factors, as already flagged, contribute to obtaining the desired result:

- pyrotechnic delays, which are inserted in the connection units and in the branches (which in practice are made with detonators having the function of receiving the impulse from one tube and transmitting it to the others);
- the (pyrotechnic) delays that can be arranged in the "final" detonators of the circuit, that is, those intended to trigger the charges.

The example shown in Figure 4.19 represents how the system works: the starter on the left (Nonel exploder/blasting machine, or another device able to trigger the shock tube); a sequence of 25 ms connection units, which in turn trigger the shock tubes of the long-delay in-the-hole detonators (500 ms). The blast lasts 675 ms and is micro-delayed, the delay being dictated by the connection units.

4.4.5   Detonating cord and relays

When detonating cord circuits are employed, it is impossible to obtain long intervals between blast holes, but only short intervals, thanks to the relays. The structure and use

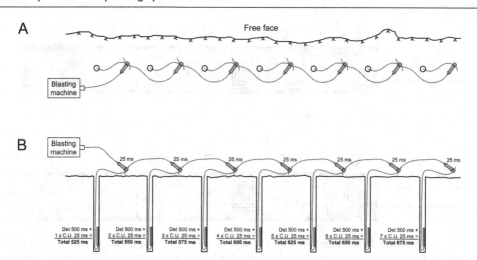

Figure 4.19 Example of timing of seven charges through the Nonel system, and representation of the total duration of the blast. A: plan view; B: cross section.

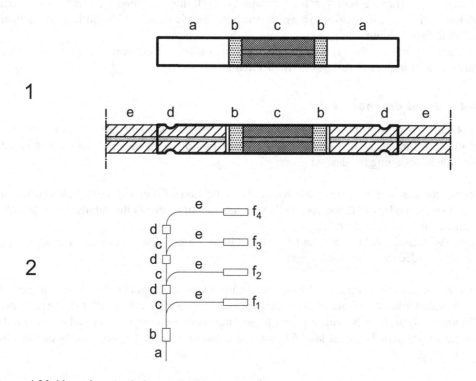

Figure 4.20 Non-electric timing systems.

1. Detonating relay (above, the device before installation; below the same, connected to the detonating cord); a: casing (aluminum tube, open at both ends); b: charges of primary detonating explosive; c: pyrotechnic delay element, similar to that of electric detonators; d: tightening on the detonating cord; e: detonating cord;

2. Example of timing of the explosions of four charges by means of detonating cord and relays; a: safety fuse; b: ordinary detonator; c: surface detonating cord; d: detonating relays; e: strands of detonating cord in the hole ("down line"); f_1, f_2, f_3, and f_4. charges whose timing is desired.

of these devices are shown in Figure 4.20. The types of common use allow the introduction of 20–30 ms delays. It should be noted that the relays contain primary explosives and must therefore be treated as detonators.

References

Agioutantis, Z., Bozinis, S., Panagiotou, G. and Kavouridis, C., 2001. Evaluation and analysis of blasting procedures for removing hard formations at the South field lignite mine, Ptolemais, Greece. 17th International Mining Congress and Exhibition of Turkey, Istanbul, TR.

Balachandar, K.G. and Thangamani, A., 2019. Studies on some of the Improvised Energetic Materials (IEMs): Detonation, blast impulse and TNT equivalence parameters. Oriental Journal of Chemistry, 35, pp. 1813–1823.

Borg, D.G., 1995. Bulk-Loaded Emulsion Explosives Technology, Coal, Chicago, p. 100.

Braithwaite, C.H., Pachman, J., Majzlik, J. and Williamson, D.M., 2012. Recalibration of the large scale gap-test to a stress scale. Propellants, Explosives, Pyrotechnics, 37(5), pp. 614–620.

Cardu, M., 2007. Two-component bulk emulsions: A revolution in the explosives manufacturing. Proceedings of the International Mining Congress and Exhibition. Turkey-IMCET, Ankara, TR, pp. 1–7.

Cavagna, G., Bartalini, E. and Locati, G., 1967. Environmental control of health hazards in a dynamite factory: Past and present. Medicina del Lavoro, 58(8/9), pp. 501–505.

Chaturvedi, S. and Dave, P.N., 2013. Review on thermal decomposition of ammonium nitrate. Journal of Energetic Materials, 31(1), pp. 1–26.

Coleman, G. and Larson, D., 1994. The chemical kiloton experiment: The utilization of 50/50 heavy ANFO at the Nevada Test Site (No. CONF-940144-). International Society of Explosives Engineers, Cleveland, OH (United States).

Cook, M.A., Farnam, H.E., Malstrom, S.D. and Peterson, W.H., 1962. Intermountain Research and Engineering Co: Booster for relatively insensitive explosives, U.S. Patent No. 3,037,452, June 5.

Dadone, P.N., 2004. Italian explosives depots from Tulps to Seveso II: A proposal for emergency planning. Probabilistic Safety Assessment and Management, Springer, London, pp. 632–637.

Dellien, I., 1982. A DSC study of the phase transformations of ammonium nitrate. Thermochimica Acta, 55(2), pp. 181–191.

Dick, R.A., 1972. The impact of blasting agents and slurries on explosives technology. IC 8560, US Bureau of Mines Information Circular, Minneapolis, MN, USA.

Dick, R.A., Fletcher, L.R. and D'Andrea, D.V., 1983. Explosives and blasting procedures manual (No. 8925). US Department of the Interior, Bureau of Mines.

Elbeblawi, M.M.A., Abdelhak Elsaghier, H.A., Mohamed Amin, M.T. and Elrawy Abdellah, W.R., 2022. Surface Mining Methods and Systems. Surface Mining Technology, Springer, Singapore, pp. 289–333.

ESSEEM Project, 2013. Training manual for rock blasting education: Initiation systems. European Shotfirer Standard Education for Enhanced Mobility, Nosrsk Forening for Fjellsprengningsteknikk (NFF, Norwegian Tunneling Society) Working Group; EFEE, Shotfiring Committee workshop, Zandvoort (Netherlands).

Fleetwood, K.G., Villaescusa, E. and Eloranta, J., 2012. Comparison of the non-ideal shock energies of sensitised and unsensitised bulk ANFO-emulsion blends in intermediate blasthole diameters. Proceedings of the Thirty-Eighth Conference on Explosives and Blasting Technique. Nashville, TN, USA.

Fodor, D., 2010. History of industrial explosive and primers production and use (II). Revista Minelor/ Mining Revue, 16(4).

Haskins, P.J. and Cook, M.D., 2007. Detonation failure in ideal and non-ideal explosives. AIP Conference Proceedings (Vol. 955, No. 1, pp. 377–380), American Institute of Physics, Batavia, IL, USA.

Hu, Q., Lu, Z. and Hua, C., 2001. Test method for impact sensitivity of insensitive high explosives (No. CNIC – 01542). China Nuclear Information Centre.

Hudaverdi, T., Guclu, E. and Kuzu, C., 2009. Application of heavy ANFO explosives in quarries nearby Istanbul. Proceedings of the 9th International Symposium on Rock Fragmentation by Blasting, Granada, Spain, pp. 111–116.

Ingman, J.S., Kearley, G.J. and Kettle, S.F., 1982. Optical and thermal studies of transitions between phases II, III and IV of ammonium nitrate. Journal of the Chemical Society, Faraday Transactions 1: Physical Chemistry in Condensed Phases, 78(6), pp. 1817–1826.

Stiher, J.F. (Ed.), 2011. ISEE Blasters' Handbook, 18th Ed. International Society of Explosives Engineers, Cleveland, OH, USA, p. 267. Library of Congress Control Number: 2010942503, ISBN 978-1-892396-19-8.

Johnson, J.N., Mader, C.L. and Goldstein, S., 1983. Performance properties of commercial explosives. Propellants, Explosives, Pyrotechnics, 8(1), pp. 8–18.

Juchem, C., Rudrapatna, S.U., Nixon, T.W. and de Graaf, R.A., 2015. Dynamic multi-coil technique (DYNAMITE) shimming for echo-planar imaging of the human brain at 7 Tesla. Neuroimage, 105, pp. 462–472.

Kane, P.G., 1979. New era in industrial explosives. The Indian Mining & Engineering Journal, 18(7).

Kholodenko, T., Ustimenko, Y., Pidkamenna, L. and Pavlychenko, A., 2014. Ecological safety of emulsion explosives uses at mining enterprises. Progressive Technologies of Coal, Coalbed Methane, and Ores Mining. Taylor & Francis Group, London, pp. 255–260.

Kim, J.W., Yoon, W., Ban, H.S., Kim, Y. and Kwon, S., 2021. Quantification of material explosive power in blasting cap test. Journal of Propulsion and Energy, 2(1), pp. 34–43.

Lee, E.A., Drake, R.C. and Richardson, J., 2014a. A view on the functioning mechanism of EBW detonators-part 1: Electrical characterisation. Journal of Physics: Conference Series, 500(19), p. 192008. IOP Publishing.

Lee, E.A., Drake, R.C. and Richardson, J., 2014b. A view on the functioning mechanism of EBW detonators-part 2: Bridge wire output. Journal of Physics: Conference Series, 500(5), p. 052024. IOP Publishing.

Lee, E.A., Drake, R.C. and Richardson, J., 2014c. A view on the functioning mechanism of EBW detonators-part 3: Explosive initiation characterisation. Journal of Physics: Conference Series, 500(18), p. 182023. IOP Publishing.

Leidig, M., Bonner, J.L., Rath, T. and Murray, D., 2010. Quantification of ground vibration differences from well-confined single-hole explosions with variable velocity of detonation. International Journal of Rock Mechanics and Mining Sciences, 47(1), pp. 42–49.

López, L.M., Sanchidrián, J.A., Segarra, P. and Ortega, M.F., 2013. Evaluation of ANFO performance with cylinder test. Rock Fragmentation by Blasting: The 10th International Symposium on Rock Fragmentation by Blasting, 2012 (Fragblast 10), CRC Press, New Delhi, India, Taylor & Francis Group, Boca Raton, FL, USA, pp. 579–586.

Mainiero, R.J. and Verakis, H.C., 2010. A century of bureau of mines/NIOSH explosives research. Proceedings, 2010 SME Annual Meeting, Phoenix, AZ.

Meyers, S. and Shanley, E.S., 1990. Industrial explosives-a brief history of their development and use. Journal of Hazardous Materials, 23(2), pp. 183–201.

Meyers, S., Singh, S.K. and Shanley, E.S., 2001. A blasting cap test for evaluating self-reactivity hazards. Process Safety Progress, 20(1), pp. 1–5.

Mohan, V.K. and Srinivas, M.A., 1980. A design criterion for permissible watergel explosives. Propellants, Explosives, Pyrotechnics, 5(2–3), pp. 93–93.

Nabiullah, M. and Singh, B., 1990. Influence of extraneous material on the impact sensitivity of explosives. Journal of Mines, Metals & Fuels, 38(7&8), pp. 141–145.

Nie, S., Deng, J. and Persson, A., 1993. The dead-pressing phenomenon in an ANFO explosive. Propellants, Explosives, Pyrotechnics, 18(2), pp. 73–76.

Nikolczuk, K., Maranda, A., Mertuszka, P., Fuławka, K., Wilk, Z. and Koślik, P., 2019. Measurements of the VOD of selected mining explosives and novel "green explosives" using the continuous method. Central European Journal of Energetic Materials, 16(3).

Oommen, C. and Jain, S.R., 1999. Ammonium nitrate: A promising rocket propellant oxidizer. Journal of Hazardous Materials, 67(3), pp. 253–281.

PECCS Project, 2019. Funded by European Commission, through the Erasmus plus programme-Pan-European Competence Certificate for Shot-firers/Blast Designers by European Federation of Explosives Engineers (EFEE), https://efee.eu/, Chapter "Explosives".

Persson, A., Jerberyd, L. and Almgren, L.Å., 1980. Testing the sensitivity of water-gel explosives to weak shocks. Propellants, Explosives, Pyrotechnics, 5(2–3), pp. 45–48.

Persson, P.A., 1980. EXTEST international study group for the standardization of the methods of testing explosives (formerly european commission for the standardization of the tests of explosives). Official Report on the 8th Meeting, EXTEST 78, held at Tatrânska Lomnica, Czechoslovakia, October 2–7, 1978. Propellants, Explosives, Pyrotechnics, 5(2–3), pp. 23–28.

Rae, P.J. and Dickson, P.M., 2019. A review of the mechanism by which exploding bridge-wire detonators function. Proceedings of the Royal Society A, 475(2227), p. 20190120.

Ringgenberg, O. and Mathieu, J., 2004. Commercial high explosives. CHIMIA International Journal for Chemistry, 58(6), pp. 390–393.

Rossetti, P., Garzia, F., Genco, N.S., Rossetti, C. and Scolari, S., 2017. Dirty bomb drones, physical-logical urban protection systems and explosive/radiological materials regulation's challenges in the age of globalization. Enhancing CBRNE Safety & Security, Proceedings of the SICC 2017 Conference: Science as the First Countermeasure for CBRNE and Cyber Threats, Springer, Germany.

Salyer, T.R., Short, M., Kiyanda, C.B., Morris, J.S. and Zimmerly, T., 2010. Effect of prill structure on detonation performance of ANFO (No. LA-UR-10-01499; LA-UR-10-1499). Los Alamos National Lab. (LANL), Los Alamos, NM (United States).

Santis, L.D., Rowland III, J.H., Viscusi, D.J. and Weslowski, M.H., 1995. The large chamber test for toxic fumes analysis of permissible explosives (No. CONF-950247-). International Society of Explosives Engineers, Cleveland, OH (United States).

Sapko, M., Rowland, J., Mainiero, R. and Zlochower, I., 2002. Chemical and physical factors that influence NOx production during blasting-exploratory study. Proceedings of the Annual Conference on Explosives and Blasting Technique, ISEE, February, vol. 2, pp. 317–330, 1999.

Shao, A.L. and Feng, S.R., 2015. Research on high-efficiency composite oil phase material for emulsion heavy ANFO. Machinery, Materials Science and Energy Engineering (ICMM 2015). Proceedings of the 3rd International Conference, Toronto, Canada, pp. 561–569.

Silva, G. and Orlandi, C.P., 2013. PANFO-A novel low-density dry bulk explosive. Performance of Explosives and New Developments. Mohatny & Singh, London, pp. 81–90.

Siskind, D.E. and Kopp, J.W., 1995. Blasting accidents in mines, a 16-year summary (No. CONF-950247-). International Society of Explosives Engineers, Cleveland, OH (United States).

Sjolin, C., 1972. Mechanism of caking of ammonium nitrate (NH4NO3) prills. Journal of Agricultural and Food Chemistry, 20(4), pp. 895–900.

SME Mining Engineering Handbook, 1996. 2nd Ed., vol. 1. Society for Mining, Metallurgy, and Exploration, Inc., 2nd printing, Library of Congress Catalogue Card Number 92–61198, ISBN 0-87335-100-2, Peter Darling Ed., p. 484.

Taguchi, T., Sasaki, S., Ariki, T. and Kimura, Y., 2005. Developments and field tests of granular emulsion explosives. Science and Technology of Energetic Materials, 66(5), p. 393.

Tsakonas, S., Jacobs, W.R. and Ali, M.R., 1979. Propeller blade pressure distribution due to loading and thickness effects. Journal of Ship Research, 23(02), pp. 89–107.

Varesh, R., 1996. Electric detonators: EBW and EFI. Propellants, Explosives, Pyrotechnics, 21(3), pp. 150–154.

Wada, Y., Yabashi, H., Tamura, M., Yoshida, T., Matsuzawa, T. and Hosoya, F., 1991. Shock sensitivity of blasting explosive cartridges. Journal of Energetic Materials, 9(1–2), pp. 105–132.

Wang, Z.S., Liu, Y.C., Zhang, J.L. and Zhang, B.M., 2003. The effects of restraint condition and charge diameter of booster dynamite HMX/F2641 on shock pressure. Explosion and Shock Waves, 23(3), pp. 248–252.

Watson, R.W., Brewer, R.L. and Mcnall, R.L., 1975. On the bullet sensitivity of commercial explosives and blasting agents. Journal of Hazardous Materials, 1(2), pp. 129–136.

Watts, R.T., 2001. Choosing the correct bulk loading equipment can help optimize your blasting program. Fragblast, 5(4), pp. 200–220.

Žganec, S., Bohanek, V. and Dobrilović, M., 2016. Influence of a primer on the velocity of detonation of ANFO and heavy ANFO blends. Central European Journal of Energetic Materials, 13(3), pp. 694–704.

Zou, D., 2017. Initiation system. Theory and Technology of Rock Excavation for Civil Engineering, Springer, Singapore, pp. 171–203.

Chapter 5

Blast holes and blasts

5.1 Introduction

A blast hole is defined by a charge placed in a cavity (hole), drilled into the rock, and equipped with the necessary triggering devices that allow it to detonate at the wanted time in order to induce fragmentation and release (or, in some cases, only one of the two effects) of a given volume of rock.

Single blast holes are rarely used in excavation techniques. In general, groups of several holes are used, detonating simultaneously (instantaneous blasts) or, more often, according to a given sequence, and cooperating in order to fragment and detach the wanted volume of rock.

A blast represents a "system", of which the single boreholes are the "components", defined by the position assigned to the charges (and therefore to the holes), by the time sequence of the explosions, and by the entity and type of the single charges.

Generally, it is preferred to use several holes instead of a single borehole, although a single, correctly designed borehole could be able to create a cavity with the desired volume. The reasons are essentially due to the following:

- to achieve better compliance with the desired geometry of the cavity to be obtained;
- to use the explosive more efficiently, where boreholes are arranged so that each one "prepares the work" for those that have to detonate later, for example, creating free surfaces that will be exploited by them;
- to reduce the "disturbance" to the rock that must remain in place after the blast.

The sketch in Figure 5.1 represents a "brutal" and a "reasoned" solution for the same problem. The latter is based on the intuitive principle of the "decomposition" of the global effect, which is the basis of any blast design. The solution is still rudimentary because it does not divide the boreholes into "functional groups" differently charged (e.g., distinguishing between "production" and "contour" holes), and is therefore improvable.

5.2 The drilling diameter

In order to place the charges in the rock according to a designed pattern, a drilling process is required. The drilling phase is very important since the rock is finely crushed following a small circular shape, a function of the bit diameter chosen with respect to the blast design. Hence, the

DOI: 10.1201/9781003241973-5

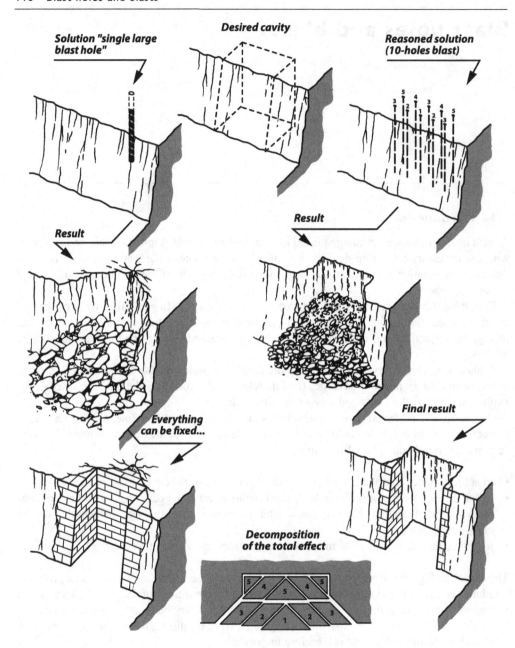

Figure 5.1 Schematic explanation of the reason for cooperation between blast holes.

drilling diameter is a very important parameter, since, from its preliminary choice, all the blast designs can be drawn.

If the holes' volume must be selected reasonably, the blast's geometry, disregarding the type, is a function of the drilling diameter.

- In fact, both burden and stemming depend on the diameter.
- By imposing a reasonable value on the utilization coefficient, that is, the charged hole length ratio to the total drilled length. This ratio, for a certain charge and a certain total length of the hole, can be satisfied by only a single diameter.
- The length/diameter ratio has an upper practical limit dictated by the need for precision: the more slender the hole, the more slender, necessarily, the drill rod: it is, therefore, less rigid, with a greater probability of deviation.

For all these reasons, even the amount of explosive that can be introduced into a hole is limited by the diameter chosen and the volume of rock that one single hole can blast.

The diameters that can be used for drilling vary from an exceptional minimum of 25 mm to an exceptional maximum of 500 mm. Of course, machines of different classes are used, from the light manual drilling rig to the heavy one. From a purely theoretical point of view, the minimum unit cost ($/m^3) of the blast would always be obtained by using the maximum diameter: in fact, on the one hand, the volumetric unit cost of the hole decreases as the diameter increases (in large-diameter drilling, coarser debris are obtained, and therefore, with the same power, more volume of the hole is produced in the unit of time), on the other hand, the large diameter makes explosives of lower unit cost-effective, and the greater entity of the charge of the single hole reduces the unit cost of loading and the incidence of initiation devices.

However, this is only valid for cases where very high production is required, obtainable with an arbitrary excavation geometry, without limits to the maximum charge to be used and to the capacity and size of the machines intended for muck removal. These ideal conditions are rarely approached: the maximum diameters are conveniently used only by the major open-cast mining units, for daily productions of tens of thousands of m^3. In the most frequent cases, a first and obvious upper limitation to the diameter follows the required production: no one would buy a very expensive machine with a high production capacity to leave it idle for most of the time. A second obvious limitation derives from the excavation geometry, which imposes upper and lower limits on the length of the holes and, therefore, indirectly on the burden and on the diameter. For example, if a bench blasting was performed by vertical holes with a height of 4 m, holes of 100 mm in diameter would not be used, which could work effectively with a burden of 3–4 m, but would be uncharged for most of their lengths due to the necessary stemming; on the other hand, if a bench blasting with a height of 30 m was carried out, a diameter of 30 mm would not be adopted, which could work well with a burden of 1–1.5 m, but could end up with a burden of less than 0.5 m or more than 2.5 m at the bottom of the bench, due to the inevitable deviations of the hole and irregularity of the face. The example shown in Figure 5.2 illustrates these abnormal solutions and, for comparison, the reasonable ones; the optimal diameter must be chosen in the range of diameters that cannot be excluded a priori.

A third limitation in the choice of diameter with reference to an optimal diameter lies in the fact that, by increasing the diameter to reduce blast costs in the view of using the volume of the holes adequately, it is necessary to increase the pattern or, similarly, decrease the drilling density (n° holes/m^2). However, the reduction of the drilling density has not only the positive effect of the reduction in the unit cost of the blast but also causes a worsening of fragmentation: greater dimensions and greater frequency of blocks requiring secondary breakage, as well as higher values of the average size of the material obtained. Consequently, an increase in the unit costs of mucking and crushing is observed for a given equipment and a given crushing plant. It can be finally stated that starting from a certain diameter, the benefit of the lower cost of the neat blast

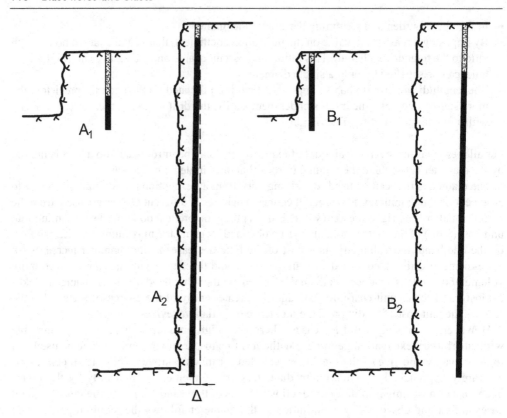

Figure 5.2 Examples of unproportioned (A_1, A_2) and well-proportioned (B_1, B_2) blast holes with regard to the relationships between diameter, bench height, and burden. In detail, in A_1, the hole is incorrectly used as the burden, and the diameter is too large in relation to the height of the bench; in A_2, the hole is well used, but the height of the bench is too large in relation to the burden and the diameter. In this last case, the inaccuracy of the bottom of the hole position (Δ) is too large in relation to the burden. The solutions B_1, and B_2, on the contrary, are well proportioned.

is cancelled. To conclude, the optimal diameter is the one that allows the attainment of the blast with minimum unit cost, as shown in the sketch of Figure 5.3.

5.3   Drilling surface, drilling density, and blasting pattern

The rock volume from a blast generally has more or less a regular prismatic shape. A free wall must be available for drill rigs, representing the surface from which the holes are drilled. The drilling density (holes/m²) is the ratio of the number of holes in the blast to the drilling surface. In the case of parallel-hole blasts, the drilling density also defines how finely the explosive is distributed in the rock; it can therefore be related to the maximum size of the fragments obtainable from one blast.

Holes can be parallel to each other, or inclined (convergent, or divergent); if they are parallel and regularly distributed, instead of the drilling density, the distribution of the explosive in the volume can be defined by providing the "blasting pattern", that is, the average spacing among the blast holes.

The drilling surfaces are almost always horizontal or vertical, as it is difficult to stabilize the machines on an inclined surface. Figure 5.4 shows different situations.

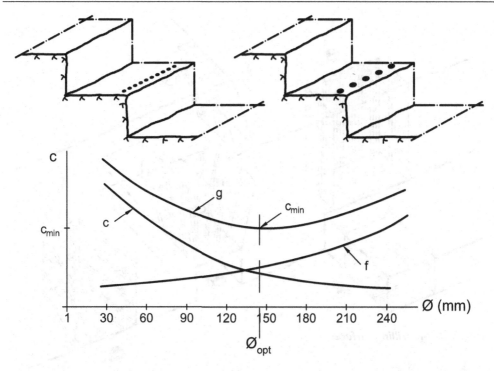

Figure 5.3 The optimal diameter for a bench blasting operation is between the two extreme solutions: A (very small diameter holes) and B (very-large-diameter holes). It is hypothesized that the blasted rock is aimed at the production of crushed stone. The diagram shows: on the vertical axis, the unit cost c (per unit of volume or mass of blasted rock) in arbitrary units and, on the horizontal axis, the drilling diameter Φ. Curve "c" represents the unit cost of the items: "explosives, detonators, and drilling", curve "f" that of items: "secondary breakage, mucking and transport to the mill, crushing"; curve "g" is the sum of the two and represents the total unit cost of production. Its minimum defines the "minimum" unit cost c_{min} and the "optimal" diameter Φ_{opt}.

5.4   Concept of functional groups of blast holes

The holes that pertain to a blast are not all in similar geometric conditions, and not all of them must induce the same effect. Consequently, they are not necessarily spaced, oriented, triggered, and charged according to the same criteria. These holes can generally be classified according to "functional groups" (Zare and Bruland, 2006). Schematically, three of them can be identified (and in many cases, a more detailed division into subclasses is possible):

- Cut (opening) holes: their function is to prepare favorable conditions for those that detonate later, creating or extending "free walls" where these are initially missing or insufficient.
- Production holes: they have to remove most of the volume, taking advantage of the favorable conditions created by the previous ones.
- Profile (or contour) holes: they must outline the contour of the wanted cavity, by removing what remains after the bulk of the work, which has been realized by the production holes (Singh and Xavier, 2005; Lu et al., 2012; Cardu, Saltarin et al., 2021; Cardu and Seccatore, 2016; Hu et al., 2014).

The functions of these groups are shown in the sketches in Figure 5.5.

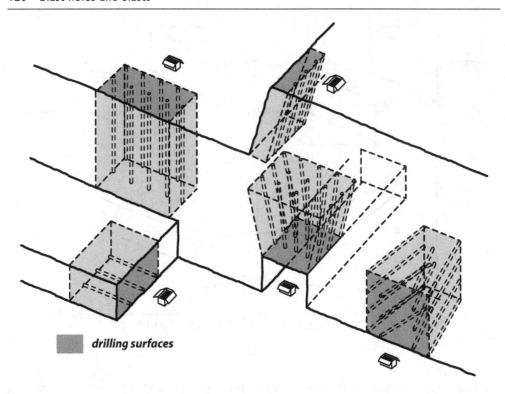

drilling surfaces

Figure 5.4 Examples of blasts with different layouts.

5.5   Types of blasts

As is well known, countless models and arrangements of blast holes are in use or have been employed, and it is inevitable that others will arise. A schematic classification is useful to show the main rules to be followed in the design.

A first distinction must be made between simultaneous blasts and blasts developed according to a certain sequence (individual detonation of each blast hole or detonation for groups of blast holes or, in some cases, two or more delayed charges in the same hole).

Simultaneous blasting is used only in special cases, either because there is usually a limitation on the amount of explosive that can be detonated simultaneously or because blasting with a suitable sequence promotes good fragmentation (Katsabanis et al., 2006). When the desired effect is essentially the subdivision of large intact elements, simultaneous blasting is adopted, because, as already mentioned, it favors the driving and the propagation of the fracture (Swift et al., 1979). The sketch in Figure 5.6 intuitively clarifies this point: when removing a wooden plank from a box, simultaneously leveraging on several points increases the probability of detaching intact planks.

This topic will be taken up again in the discussion of "splitting" technique.

In the next paragraphs, sequential blasts are analyzed, which are the most common, and are essentially aimed at achieving a good fragmentation.

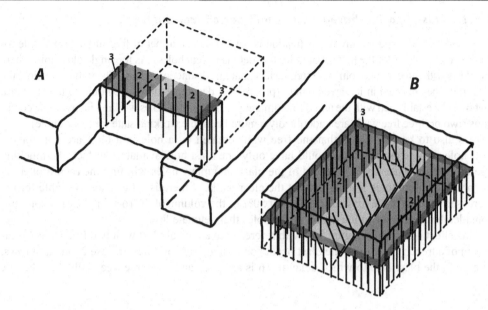

Figure 5.5 A (excavation of a trench): the cut holes (group 1) have the function of opening a vertical groove, where the walls can be used as free surfaces for the production holes (group 2), which progressively enlarge the cavity. The contour holes (group 3) "refine" the sidewalls of the excavation.

B (deepening of a pit, starting from a horizontal free surface): the opening is performed by the row of paired inclined holes (group 1). They open a groove that will be used as a free surface by the vertical production holes (group 2), and finally, the profile holes (group 3) provide the contour.

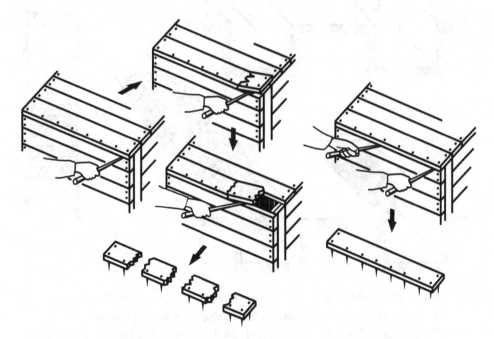

Figure 5.6 Analogy showing the different fragmentation and detachment effects resulting from sequential and simultaneous blasting.

5.5.1   Classification based on the number of free surfaces

The most common and intuitive classification is based on the number of free surfaces available for the first blast hole or for the first set of holes that detonates (cut holes). As for single blast holes, there are blasts with three, two, or only one free surface, but this distinction refers to the initial situation: the blast must be analyzed in its development (progressive decomposition), according to the sequence that must be established with the aim of maximizing the number of holes working in favorable conditions (two or three free walls are undoubtedly a more favorable working state for a blast hole).

It is also to be considered that the free walls created by a hole in a blast are not, for the next hole, as free as those that would have only one blast hole available: in fact, the mucking operation takes place only at the end of the blast, and many holes will find the rock broken by the previous blast holes leaning against the clear walls. If there is little space available for the increased volume of the blasted rock (about 50% of the volume of the rock in place), it can offer considerable resistance to the action of the holes that detonate last.

This effect is shown in Figure 5.7, with reference to a typical blast with two free walls (excavation of a trench): in case 1, the blast occurs on a single row of holes; in case 2, on eight rows. In case 2, the blast hole to which delay n. 16 is assigned and has three free walls, like the one

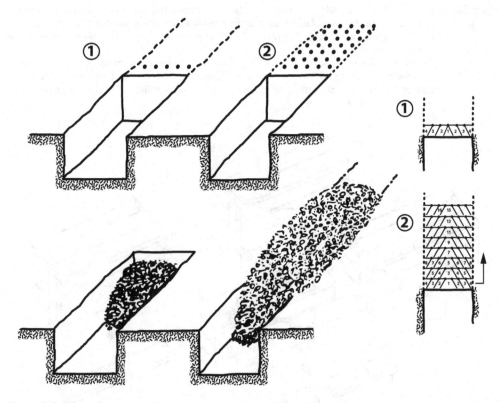

Figure 5.7 Blasts for a trench excavation can be organized on one or many rows.

  1. Perspective view (left) and plan (right) of a single-row blasting scheme.
  2. Perspective view (left) and plan (right) of a multi-row blasting scheme.

having delay n. 2, but it works with considerable difficulty due to the heap of blasted rock ahead. In practice, in such cases, the powder factor (kg of explosive/m³ of on-site rock volume) of the blast holes of the rows following the first has to be increased by 20–30%.

The numbers on the plan views show the timing of blasts; the arrow shows the backward direction of the face due to the multi-row blasting.

5.5.2   Blasts with three free surfaces

When three free walls are available, assigning an individual delay to each blast hole can be a favorable circumstance. The hole closest to the vertex of the trihedron detonates first, allowing the presence of three free walls for the one that detonates afterward, and so on. In this case, there are no holes with opening functions. This condition occurs quite often in both civil and mining open-cast works (e.g., widening of roads, exploitation of benches in the quarries) and less often underground (e.g., widening of large underground cavities, down warding of mining exploitations).

Some of the many possible examples are shown in the sketches of Figure 5.8.

The numbers show the timing of explosions. In case 3, the number zero shows a row of contour holes that are triggered simultaneously before the production blast (presplitting).

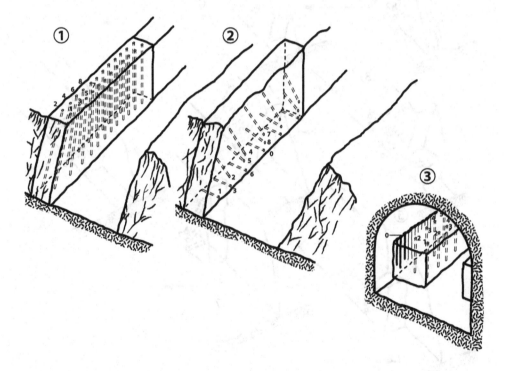

Figure 5.8 Examples of blasts with three free walls.

1. Widening of a trench, with vertical holes.
2. Widening of a trench, with holes arranged according to vertical fans, drilled and charged from the floor.
3. Enlargement of an underground cavity.

5.5.3   Further explanations about the concept of decomposition of a blast

The "decomposition" has to be understood according to a chronological meaning (Seccatore et al., 2015; Cardu et al., 2015) and should consist of regressing the geometry of the face after each detonation, *erasing* the rock already broken. It is impossible to actually perform it on an ordinary blast (even an ultra-fast film shooting can only take up the external aspect of the blast under development), but it can easily be dealt with conceptual planning: in this case, it is just necessary to assign a reasonable blast radius to each hole, and progressively redesign the face, erasing at each detonation what has been theoretically removed. This is, therefore, an idealized reconstruction of what should happen, that is, an option of checking whether a project is reasonable or not. Of course, the hypothesis implying that surfaces progressively released are flat and intersect on the axes of the holes is an idealization of the real phenomenon. Figure 5.9 shows a theoretical decomposition of blast 2 with reference to Figure 5.8.

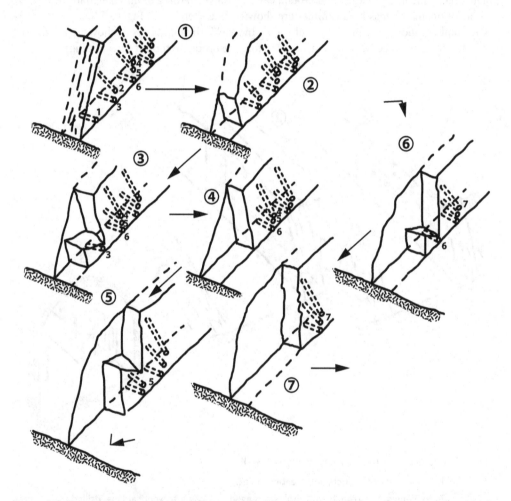

Figure 5.9 Decomposition of the effect of the blast by vertical fans for a trench widening (according to sketch 2, Figure 5.8). The numbers show the chronological succession of the face configuration changes due to the consequent explosions. These configurations are represented by ideally removing the heap of debris that is gradually obtained.

5.5.4 Blasts with two free surfaces

It is the most common case in both open pit and underground excavations (excluding the blind-bottom progressions). When two free walls are available, a positive dihedral is obtained, one face of which is assigned the role of drilling surface.

The basic principle when designing these blasts is very simple: a hole (or a group of holes), called an opening, is assigned the task of creating a new free surface, that is, a groove on one of the faces of the dihedral (and specifically the one that does not represent the drilling surface); the detonation of the opening hole (or holes) therefore leads back to the case already seen, relating to the three free surfaces.

Some typical examples are schematically illustrated in Figures 5.10 and 5.11, where the numbers refer to the timing of the explosions; the sketches show the trace of the volume that (theoretically) belongs to the opening holes.

In some sketches, it has been hypothesized that the timing is sufficient to ensure that the rock removed from one blast hole has already undergone a displacement when the next one detonates: this, only for reasons of conceptual clarification of the behavior of the blast, since, if the timing is organized according to very short delays, the volume that belongs to one hole may be already fragmented, without having undergone significant displacements yet, when the next hole detonates.

Figure 5.10 presents some examples of blasts with two free surfaces: in A, a bench blast with vertical holes on a single row is schematized, and in B, a similar blast with two rows of holes. In these cases, the drilling surface is the upper tread of the bench. In C, a bench blast with horizontal holes is represented: in this case, the drilling surface is the riser of the bench. This scheme is used only for particular reasons, for example, the limited height of the bench or the hard access of the drilling machines to the upper tread, making the exploitation with vertical holes inadvisable. A result with irregular and uncontrollable walls can be obtained using this scheme. In D, a reverse bench blast with horizontal holes for underground excavation (now rarely used) is schematized: the drilling surface is, in this case, the elevation of the bench, and the drillers have to climb the pile of rock accumulated from the previous blast to drill the holes. Figure 5.11 proposes other examples of blasts with two free surfaces: in E, a ring blast for rock exploitation in large underground stopes is schematized: the drilling surface is, in this case, the wall of a tunnel leading into the stope itself. In F, a half-ring blast for the expansion from d_1 to d_2 of the width of a tunnel is represented. In G, the same problem is solved with a blast with parallel holes (preferable solution when a regular outline is desired). A blast for the enlargement of a rectangular stope is highlighted in H.

5.5.5 Blasts with one free surface

This case systematically arises in blind-bottom excavations (tunnels, shafts, and raises): the drilling surface coincides with the free surface, which is completely delimited by negative dihedrals. The blasts in this category are the most complex. For a (non-exhaustive) description of the related schemes, refer to Mancini et al. (1998), where about 200 examples were collected and analyzed. Some typical schemes are shown in Figures 5.12 and 5.13.

The opening, that is, the creation of free walls that allow the "production holes" to operate in geometrically favorable conditions, can essentially be obtained according to the following ways:

1) With one or more blast holes perpendicular to the face, having the function of opening a crater: these blasts are rarely used, except in the case of very narrow cross sections (Hino, 1956, 1959).
2) With a group of holes inclined with respect to the face: the most commonly adopted, especially in the excavation of tunnels or shafts with large cross sections, is the V-cut blasting

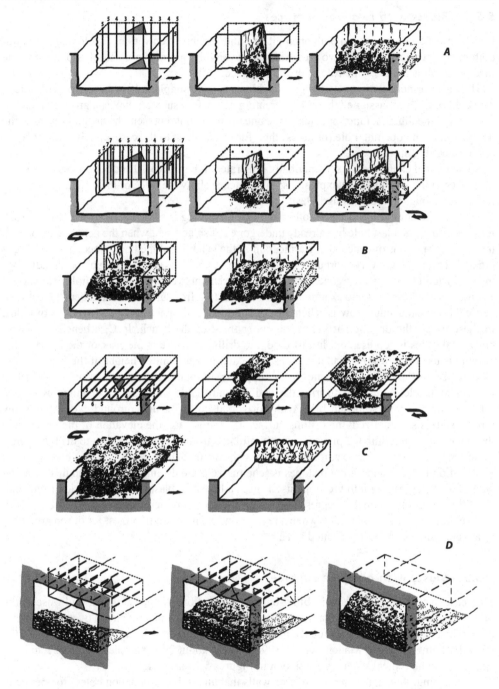

Figure 5.10 Examples of blasts with two free surfaces.

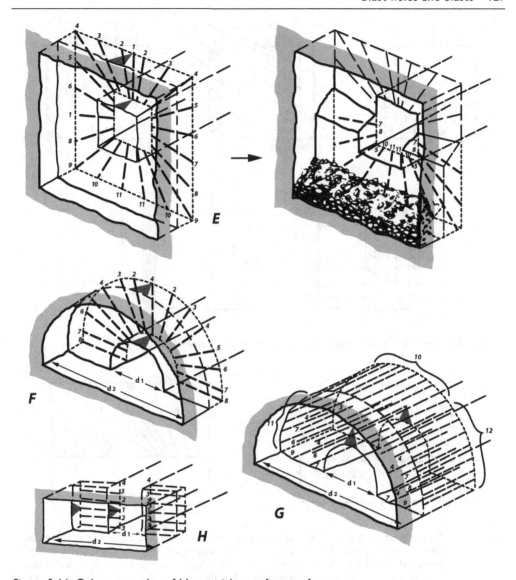

Figure 5.11 Other examples of blasts with two free surfaces.

(Berta, 1994; Lee et al., 2005; Kim et al., 2007; Konya et al., 2017; Xinkuan et al., 2018; Lou et al., 2020; Alipour and Mokhtarian, 2021).

3) With mechanical + explosive mixed excavation system, that is, providing reduced free surfaces by means of holes perpendicular to the face, often reamed (dummy hole/s), and progressively widening them with parallel holes: parallel-hole cuts, burn cuts (Hagan, 1984; Johansen and Mathiesen, 2000; Allen and Worsey, 2015; Seccatore and Cardu, 2018; Godio and Seccatore, 2019; Desai and Badal, 2000).

In any case, the goal is to minimize the number of holes in the blast that operate in the most unfavorable condition. Figure 5.14 schematically shows a very commonly used V-cut blast.

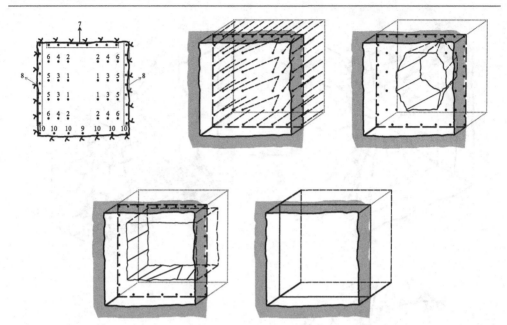

Figure 5.12 Blasting pattern with one free surface, inclined holes cut.

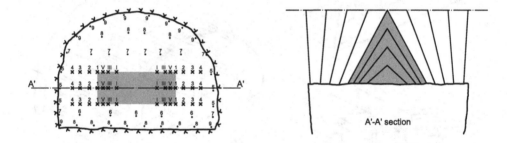

Figure 5.13 Example of a blast in an average cross-section (60 m²) civil tunnel in granite (Sweden), with multiple V-cuts, drilled with a wagon drill. Front view (left) and longitudinal section (right). Powder factor: 1.0 kg/m³ (straight dynamite and nitrate
explosive), average drilling density: 1.4 holes/m², theoretical pull: 4.6 m, actual
pull 4.4 m. The numbers show the delay sequence (the Roman numbers refer to
the short delays).

5.5.6   Splitting blasts

The splitting blasts are not required to induce the breakage into the rock, but only to create one
(or more) split fracture, isolating a volume of rock to be removed or fragmented. Two main
cases can be identified:

- Pre-splitting: the fracture aims to isolate a volume of rock to be subsequently blasted and
  removed (Worsey et al., 1981; Zhang et al., 2020). This may be necessary to avoid damage to

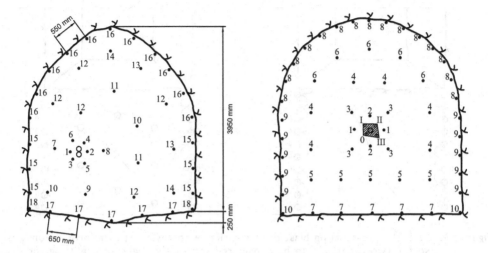

Figure 5.14 Examples of blasts for tunneling, with parallel holes cut.

A. Small cross-section (13 m²) blast: hydraulic tunnel in gneiss and granite (Finland) with a "Coromant" (eccentric) opening, drilled by a wagon drill. Powder Factor: 2.1 kg/m³ (ANFO), average drilling density: 3 holes/m², theoretical pull: 3.6 m, and actual pull: 3.2 m. The numbers show the order of detonation.

B. Blast for a hydraulic, medium cross-section tunnel in peridotite (Italy), Parallel holes (spiral) cut, drilled by a wagon drill. Powder Factor: 3 kg/m³ (dynamite and contour charges), average drilling density: 2.7 holes/m², theoretical pull: 3.2 m, and actual pull: 3 m. The numbers show the detonation order, and the Roman numbers refer to the short delays. The cut area is dashed.

the rock that must remain in place (Yang et al., 2021), to limit the transmission of vibrations induced by the next blast (Ma et al., 2021), or for both reasons (Yang et al., 2019).

• Exploitation of dimension stones: the fracture aims to isolate intact blocks, which will be removed and squared according to the wanted size, and traded.

In the first case, the development of some fractures besides the planned one is tolerated, in the second this must be excluded. In both cases, the blast consists of an alignment of parallel holes, with small and regular spacing (usually ranged between 6 and 25 times the diameter), weakly loaded (with strongly decoupled charges or with explosives with very low disruptive power and low detonation pressure), lying straight on the desired parting plane, which are detonated simultaneously.

The pre-splitting can be combined as contour holes in an ordinary blast (in this case, the contour holes detonate first) as reported in the sketch of Figure 5.15 (left).

The blasts for the exploitation of dimension stones require at least three free walls available, to allow a given displacement without inducing any damage to the blocks (see Figure 5.16).

5.6 The blasting plan

The goal of the blasting plan is to manage the blasting project correctly; it proposes the position, diameter, length, and orientation of the holes, the type and amount of explosives, the initiation devices, the amount of stem, and the timing sequence. Even in systematic blasting operations,

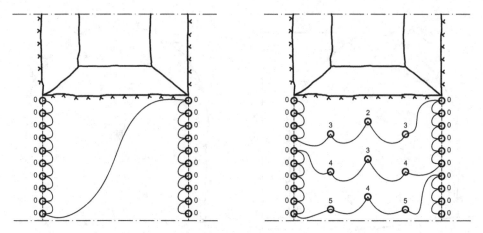

Figure 5.15 Left: the pre-splitting blast occurs earlier with respect to the production blast, so the two events are to be considered separately. Right: pre-splitting and production blast are charged and timed according to a sequence in which the pre-splitting holes detonate first.

Figure 5.16 Splitting sequence starting from a bench with three free surfaces. From left to right: displacement from the orebody of large volumes of rock with a prismatic, regular, geometry; splitting into slices and tipping on the yard.

the pattern has to be open to small changes dictated by local circumstances. However, it must not be "open" to uncontrolled inaccuracies (Mandal et al., 2008; Bakhtavar et al., 2017).

Working on rocks with heterogeneous materials unknown in local details and on irregular surfaces makes it impossible to respect strict geometric tolerances as in artificial materials; however, reasonable tolerances must be established and observed (Cardu, Godio et al., 2021).

The blasting plan can be summarized in two different documents. A graphic and quoted representation drawn on an adequate scale is commonly used for representing the blast. The data relating to the charges/hole adopted and the blasting sequences are instead reported in a table (also called "Charging table"), attached to the graphic representation. In case of complicated situations, the blasting plan should also include a third document that reports the longitudinal sections of the blast holes.

The example provided in Figure 5.17 relates to a blast in a tunnel, with parallel holes cut, 4 m pull, 56 holes (of which 54 charged, with a diameter of 51 mm, and 2 dummy holes with a diameter of 102 mm).

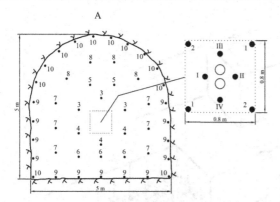

A

B

# of charges and function	# of cartridges per charge	Explosive per charge
8 - cut holes delays I, II, II, IV, 1, 2	4 GD1 + 4 Tutagex	5.2 kg
21 - stoping holes delays 3...8	3 GD1 + 4 Tutagex	4.5 kg
7 - floor holes delays 9, 10	4 GD1 + 4 Tutagex	5.2 kg
18 - profile holes delays 9,10	1 Tutagex + 5 Emuldin + 1 Tutagex	2.45 kg

Figure 5.17 Example of a plan for tunnel driving.

A. Graphical representation, in a front view, of the arrangement of the holes of the whole blast. Separately, and in a different scale, the front view of the cut, where the sequence of timing is shown;

B. Charging table.

5.7   Quality of the result

The goal (apart from the case where only a splitting effect is wanted) is to induce the fragmentation of a given volume of rock on site according to a predetermined size.

The minimum to be expected is that the blast reduces a given volume of rock to a transportable size, with a tolerable percentage of oversizing; the maximum (the "perfection") is that it detaches and reduces to fragments with a specific particle size distribution only for the volume assigned by the plane, leaving a cavity with regular, stable walls and, obviously, corresponding to the design geometry (Figure 5.18). This "perfection" is practically never reached, but it can be used as a goal to judge the quality of the result, which is significantly enhanced as the "perfect" result is progressively approached. For this purpose, it is obviously necessary to use quantitative indicators, as well as establish criteria to distinguish that part of the imperfection that is due to the characteristics of the medium being excavated from that which is due to design or execution errors (which can therefore be improved by operators).

The main points that have to be taken into account for estimating the quality of a blast are

- the compliance of the new rock profile with the planned excavation contour;
- the state of the profile walls;
- the fragmentation of the muck.

All these aspects are deepened in the following.

5.7.1   The compliance with the planned excavation contour: bench blasting

The ideal result would be the coincidence of the plane where the contour holes lie with the residual wall, as shown in Sketch I of Figure 5.19. This rarely occurs: ordinary blasts

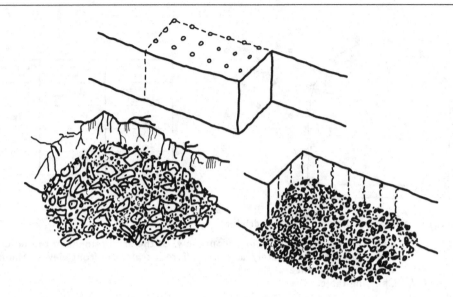

Figure 5.18 Sketch of a bench blasting ready for triggering (top) and possible result: poor (bottom, left) and excellent (bottom, right).

typically damage more rock than predicted by design, and even when the cubic meters match what was calculated, they are seldom distributed as planned. Apart from any errors due to the positioning of the drill and to the holes' deviation (inaccuracies that obviously are not attributed to the blast), the most frequent discrepancies observed between the real and the ideal result are (sketch II, Figure 5.19):

- backward of the edge of the residual bench with respect to the alignment of the mouths of the contour holes;
- ledges and recesses, with respect to the alignment of the holes, of the bench foot;
- vertical profiles of the bench, not straight, as the design would like, but with recesses and ledges.

The main damage resulting from these inaccuracies is that they make difficult or impossible the correct setting of the next blast. At the end of the work, the damage consists of failure to comply with the agreed scheme.

Checking the magnitude of these inaccuracies naturally involves detecting (once cleaned) the edge and foot of the bench and a number of vertical profiles of the remaining face, regularly spaced and in a statistically significant number to rebuild an average profile (Konya and Walter, 1991). The results of these surveys must be compared with the project to calculate the volume exploited in excess (overbreak) with respect to the design geometry. From the survey and comparison, two useful indicators of the accuracy of the result can be obtained. Both indicators are expressed in meters:

- the average overbreak (OB), which is the ratio between the volume removed in excess and the face area (m^3/m^2) and can also be expressed as the ratio between the surface (blasted in excess and the perimeter of the contour holes (m^2/m);
- the average back-break (RS), which is the average retreat of the edge of the bench compared to the alignment of the contour holes (m^2).

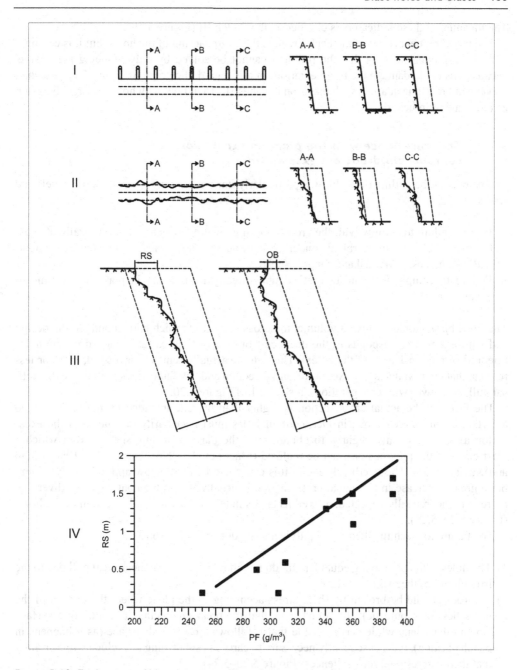

Figure 5.19 Evaluation of the accuracy of bench blasting with sub-vertical holes.

    I. Plan view (left) and vertical sections (right) of a blast that achieved a perfect result.
    II. Plan view (left) and vertical sections (right) of a blast with significant inaccuracies.
    III. Definitions of back-break (RS) and average overbreak (OB).
    IV. Correlation between the back-break and the powder factor from surveys in limestone quarries.

The meaning of these indicators is clarified in the sketch III (Figure 5.19).

The most frequent cause of the back-break is the overcharging of the holes, but it is certainly not the only one, meaning that the problem cannot be solved by only reducing the charge, because this could damage the fragmentation. In this regard, the graph of Figure 5.19, showing surveys in limestone quarries with blasts on benches 10–15 m high with holes of large diameter, can give a first indication.

5.7.2   The compliance with the planned excavation contour: blind-bottom excavation

The respect of the outline of the blast is particularly important in the excavation of tunnels and shafts:

- In fact, failure to comply with the cross-section geometry (which is systematically affected by inaccuracies due to overbreak, underbreak, or out of shape) leads to an increase in support and coating costs (Mandal and Singh, 2009).
- Failure to comply with the theoretical pull directly leads to an increase in the time of excavation.

The ideal blast should remove a volume, the cross section of which corresponds to the design and whose length corresponds to the distance from the drilling surface reached by the holes (theoretical pull, or length of the holes). The volume actually removed is a solid, more or less regular, that approximates "by excess" the cross section and "by fault" the length of the theoretical pull, as shown (with exaggeration) in sketch 1 of Figure 5.20.

The fact that the actual cross section is higher than the one intended by the project is, to a certain extent, unavoidable; in fact, contour holes must necessarily deviate from the excavation axis, even if only slightly, to obtain, after the blast, a drilling surface from which a blast equal to the previous one can be realized (Maerz et al., 1996; Ibarra et al., 1996; Singh and Xavier, 2005). The overbreak due to this cause can be kept below approximately 15 cm, but a greater increase in the actual cross section is usually due to the contour holes diverging more than theoretically expected, as well as errors in the positioning and direction of the holes (Figure 5.20–5.23).

The failure to reach the theoretical pull is mainly due to two reasons:

1) The holes, although having equal length, do not all reach the same theoretical pull due to the irregularity of the drilling surface.
2) The rock should be broken by shear stress according to the plane where the bottom of the holes lies (or should lie). It is impossible that the breakage occurs according to a surface beyond that plane while it is probable that it follows a backward surface (as it happens in bench blasting). It is consequently necessary to carry out subdrilling to exclude with certainty that there are residual rock saliences (Figure 5.20–5.24).

The compliance with the excavation contour involves the detection of numerous cross-section profiles of the excavation (in this case, more than one per blast), to obtain an effective average profile to be compared with the project. The average overbreak (OB) thickness is an effective synthetic indicator of the goodness of the result (In the case of tunnels, the floor from the contour is excluded for practical reasons, considering only the roof and walls, as the floor is normally brought to the desired level with carryovers; see Figure 5.20–5.25).

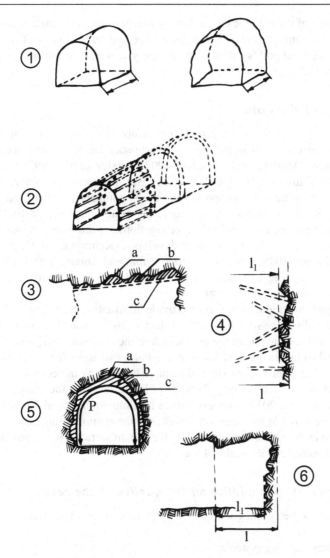

Figure 5.20 Assessment of the geometric accuracy of a blast in a tunnel.

1. Theoretical volume to be removed (left) and volume actually removed by the blast (right).
2. An amount of volume removed in excess compared to the necessary is unavoidable since some "lookout" is to be applied to the contour holes.
3. Section carried out on a perimeter hole, in which the actual outline of the (a) excavation, (b) the overbreak, and (c) the position of the hole representing the desired theoretical outline are highlighted.
4. Vertical section of the face after the blast, highlighting the positions of the ends of the holes and the actual surface, compared to the theoretical one. l is the theoretical pull; l_1 is the one actually achieved.
5. Actual contour of (a) the excavation and (b) section of the outside contour compared to (c) the outline of the project; the outside contour is usually expressed in m, by the ratio of the area blasted in excess b to the perimeter of the contour P (this definition lends better to define the overbreak in case of blind-bottom excavations).
6. The "efficiency" of the blast is expressed by the ratio (%) of the theoretical pull l to that actually reached l_1.

As for the respect of the pull, the evaluation obviously requires the measurement of the actual pull of a significant number of blasts to obtain an average reliable value. The ratio (%) of this average to the theoretical pull is called blast efficiency, and it is an indicator of the goodness of the result (Figure 5.20–5.26).

5.7.3   Quality of the walls

The blast inevitably also stresses the rock that is not intended to be removed. It can, for instance, remove part of it (as in the case of overbreak), but it can also weaken it, without inducing a collapse (Martino and Chandler, 2004; Emsley et al., 1997; Silva et al., 2019; Raina et al., 2000; Ramulu et al., 2009). In most cases, the rock is not intact before the blast, but verifying that its state does not worsen too much compared to the original conditions is necessary. The potential presence and the thickness of any damaged portion adjacent to the residual wall can be detected by seismic investigations, as the sound velocity in the rock is significantly lowered if fractures have been developed (Kahraman, 2002; Paillet and White, 1982; Grady and Kipp, 1979; Fjaer et al., 1989; Rubin and Ahrens, 1991). The visual inspection instead gives very vague indications because the term of comparison (i.e., the original "intact rock") is obviously not observable.

A very useful parameter able to indirectly provide information pertaining to the regularity and the quality of the walls is the so-called half-cast factor (HCF). The HCF is easy to detect, and it can be computed as the ratio (%) of the total length of the remaining half casts (half barrels) left on the rock after the blast to the total length of the holes that were drilled (Figure 5.21).

It is an indirect indicator of the extent of damage induced by the contour blast holes (Singh and Lamond, 1993), as what directly shows is only the fact that the fracture has been "well driven" (Dey and Murthy, 2012). However, since it is known that good fracture driving is only achieved if the pressures in the holes do not exceed the minimum necessary for detachment, HCF can be taken as an indicator of severe (HCF zero), moderate (HCF medium), or low (HCF high) mechanical damage to the residual wall.

5.7.4   Influence of rock quality on the quality of the result

Achieving correct geometry and good-quality walls depends on three factors:

- drilling accuracy (drilling system);
- suitable and accurate loading and timing (explosive system);
- quality of the rock mass.

The last factor is the only one not under the control of the operator. However, its influence must be examined because, in order to decide whether a result is acceptable or not, it is not enough to describe the rock mass as "bad", "average", or "good" (for this purpose, the indicators described above, RS, OB, HCF, or others that the practice can suggest may be used) but it is also necessary to find a criterion to determine whether, and to what extent, it can be improved (Cardu, Godio et al., 2021).

Rock quality indicators have been developed from rock mechanics studies related to stability problems but are also appropriate to assess the quality of the expected result for excavation purposes (Rehman et al., 2018).

In the following examples, the rock mass rating (RMR) is used to indicate rock quality. The graphs in Figure 5.22 show the values of OB and HCF from a series of tunnels realized in rocks of different quality (different RMR values), using a computerized drilling system in all cases (which guarantees a centimetric accuracy in the positioning of the holes and therefore allows to

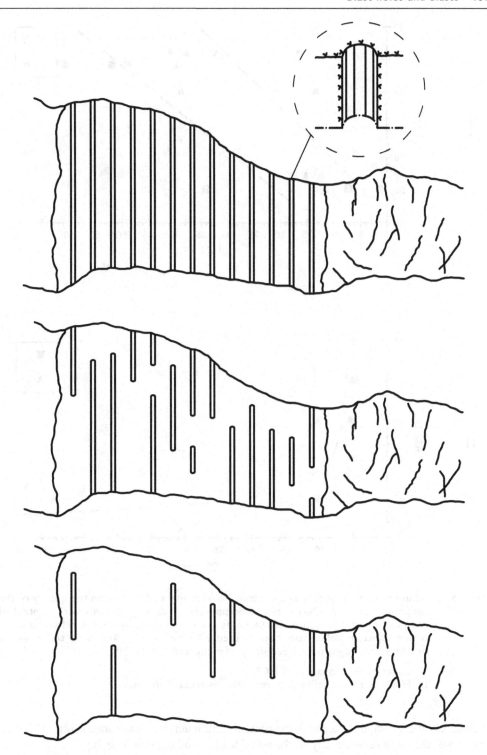

Figure 5.21 Illustration of the concept of HCF (Half-Cast Factor). From top to bottom, residual face examples with excellent (HCF~100%), medium (HCF ~50%), and poor (HCF ~ 10%) result.

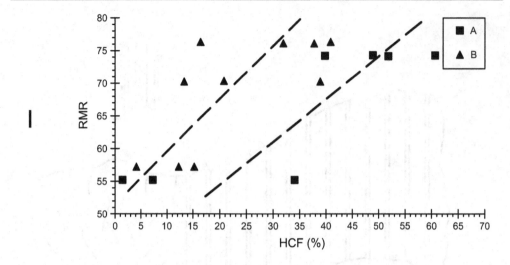

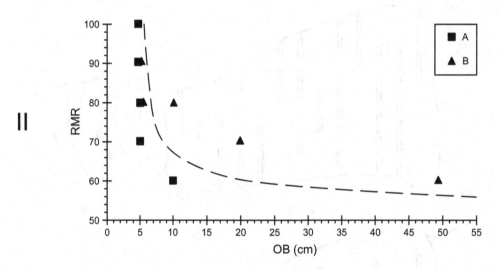

Figure 5.22 Influence of rock quality and contour hole loading techniques on the quality of the result in tunneling. Only cases of computerized drilling systems with control of the position and direction of the holes are considered. Symbol A refers to cases where contour holes have been loaded with decoupling, and B to those where ordinary cartridges have been used (Mancini and Cardu, 2001).

    I. Effect of rock quality on HCF.
   II. Effect of rock quality on overbreak, expressed in cm.

exclude that the non-compliance with the contour is attributable to inaccurate drilling). In such conditions, the result depends only on the holes' loading and the rock's quality.

The OB values shown in the graphs are net to the part contributed by the unavoidable divergence of the contour holes. Regarding loading, the cases in which the contour holes were loaded with ordinary cartridges (complete coupling) and those in which they were loaded with thin pro-filing charges are indicated with different symbols. Despite the dispersion, both, the usefulness of loading the contour holes with special charges, and the considerable influence of the quality of the rock, are evident in the result.

5.7.5 Compliance with fragmentation requirements

The blasted rock is not required to comply with a certain particle size distribution firmly; how-ever, at least a given "grain size range" must be respected.

Usually, the required fragmentation is fine, but there are frequent cases where the opposite need exists. In most cases, the control of this aspect is purely visual and qualitative.

In the systematic production blast, an indicator commonly used for comparing the results of different blasts is given by the frequency of secondary breaking operations (reduction of oversized blocks with various systems), on the yard or at the mouth of the primary crusher (when the material is intended for crushing). Similarly, when the material has to be crushed, a commonly used practical indicator is the percentage of "primary fine material", which is what the grid of the plant head excludes from the primary crushing, because it is already quite small in size (Figure 5.23).

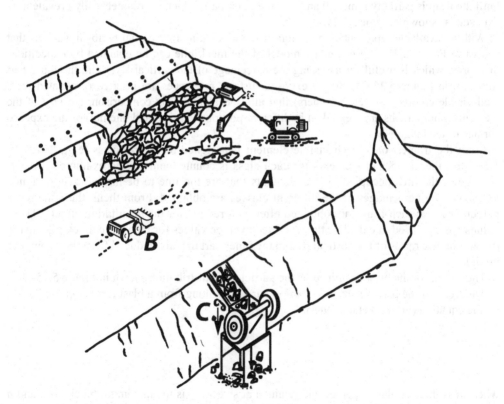

Figure 5.23 Operations and characteristics of the equipment on which fragmentation has a noticeable influence: (A) need of secondary breaking, (B) mucking (bucket capac-ity of the shovel), and (C) crushing (percentage of "primary fine materials" and crusher characteristics).

A reliable particle size analysis of the blast provides more comprehensive information, but it is laborious to obtain; of course, it is practically impossible to sieve the product of a blast, but quite good results can be obtained thanks to photographic methods (Swithenbank et al., 1976; Van Aswegen and Cunningham, 1986; Hunter et al., 1990; Maerz et al., 1996; Nefis and Talhi, 2016; Engin et al., 2020; Shehu et al., 2020; Kawalec et al., 2019; Sanchidrián and Singh, 2012).

It is necessary to point out that the visual examination of the muck pile suggests an abundance of fragments much larger than the real ones. This mistake is due to two main reasons: on one side, the larger elements hide the finer ones, and on the other side, a "stratification" or "segregation" of the muck pile naturally occurs. Combining these two phenomena shows that the surface layer is much richer in large blocks than the inner portion. Deepening the aspects that generate this heterogeneity, two reasons can be recognized:

• Fine fragments can infiltrate larger fragments.
• The surface layer is predominantly made of rock falling from the top of the bench, where the specific charge (g/m^3) is lower, the holes being filled with stemming, so that the fragmentation is coarser.

In order to have a correct idea of the actual fragmentation, it is therefore necessary to wait until the heap is partially removed and that the inner part (which is volumetrically prevalent) is exposed, as shown in Figure 5.24.

With an available representative section of the muck pile after partial removal, such as that shown in Figure 5.25-A, a geometric model of the muck pile can be obtained by expeditious measures, which is useful for assessing the percentage of the cortical layer and the inner part (diagram in Figure 5.25-C). On the exposed surface, references are inserted (graduated rods, well-visible colored spheres, or others) that allow evaluating, by comparison, the size of the elements; photographs are then taken in different portions (cortical and internal) of the exposed surface of the pile.

From the frames, once each visible element is assigned to its size class (e.g., 2–1.5 m; 1.5–1 m), the number of elements for each class is counted and having evaluated the percentage of the area occupied by the elements that are too fine to be individually assigned to a class, the percentages of the different classes are obtained. From them, the volumetric percentages for each frame are obtained after a correction for the occultation effects. These values are then used to calculate the weighted average values for the whole muck pile, using the volumetric ratios of the cortical part to the inner part that are deduced from the geometric model.

The result, for the blast examined, is the particle size distribution shown in Figure 5.25-D.

The trend of the particle size distribution curves resulting from a blast can be approximated by the equation (Fornaro et al., 1989):

$$R = e^{-a\left[\frac{x}{(x_m - x)}\right]^b}$$

where R is the cumulative percentage retained at size x, x_m is the maximum block size, and a and b are experimental constants characterizing individual distributions (b is generally little less

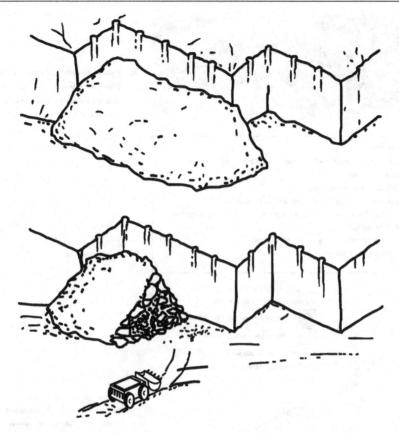

Figure 5.24 Muck pile segregation effect: only after partial removal a correct idea of fragmentation is obtainable, which is initially distorted by the prevalence of large blocks in the surface layer.

than 1, and *a* varies in a wider range, being mainly influenced by the specific charge). In the case examined, the equation that best approximates the distribution is:

$$R = e^{-1.29\left[\frac{x}{(2-x)}\right]^{0.87}}.$$

The data of the blast are provided in Figure 5.25-B. It has to be noticed that the powder factor was very low, as a coarse fragmentation was wanted (d_{50}, in fact, is 0.7 m).

5.7.6 Conclusions on performance indicators

The indicators examined make it possible to objectively compare (on the basis of data and not personal impressions) the different techniques applied in the same medium. Of course, depending on the type of work considered, different factors can be of great or marginal importance. For

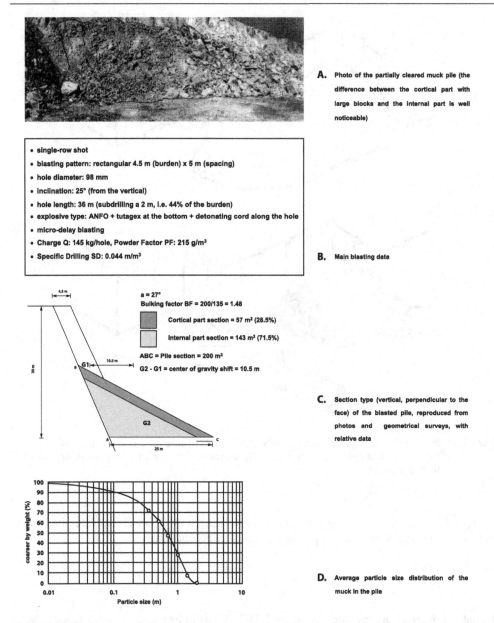

A. Photo of the partially cleared muck pile (the difference between the cortical part with large blocks and the internal part is well noticeable)

- single-row shot
- blasting pattern: rectangular 4.5 m (burden) x 5 m (spacing)
- hole diameter: 98 mm
- inclination: 25° (from the vertical)
- hole length: 36 m (subdrilling a 2 m, i.e. 44% of the burden)
- explosive type: ANFO + tutagex at the bottom + detonating cord along the hole
- micro-delay blasting
- Charge Q: 145 kg/hole, Powder Factor PF: 215 g/m³
- Specific Drilling SD: 0.044 m/m³

B. Main blasting data

a = 27°
Bulking factor BF = 200/135 = 1.48

Cortical part section = 57 m² (28.5%)

Internal part section = 143 m² (71.5%)

ABC = Pile section = 200 m²
G2 - G1 = center of gravity shift = 10.5 m

C. Section type (vertical, perpendicular to the face) of the blasted pile, reproduced from photos and geometrical surveys, with relative data

D. Average particle size distribution of the muck in the pile

Figure 5.25 Evaluation of the fragmentation obtained from a blast in a limestone quarry.

example, in an excavation with mining purposes, out-of-shape and overbreak, which are unacceptable in a civil excavation, can be accepted.

In the summary of Table 5.1, the indicators of "success" are collected, considering separately three categories (civil, open-cast, and underground mining) and showing for each the most common practical reasons for control.

Table 5.1 Main indicators of the quality of the blast result for three different categories of works

Type of work	Aspect to consider	Main reasons	Indicators
Civil	Excavation geometry	Terms and conditions	OB
	Quality of the walls	Terms and conditions	HCF
	Fragmentation	Ease of mucking	D_{max}, D_{50}
Open-cast mining	Excavation geometry	Regularity of the process	RS, OB
	Quality of the final walls	Stability of the final walls	OB, HCF
	Fragmentation	Ease of subsequent operations (mucking and crushing)	Frequency of secondary breakage D_{max}, D_{50}
Underground mining	Excavation geometry	Stability, selectivity	OB
	Quality of the final walls	Stability	OB, HCF
	Fragmentation	Ease of mucking	D_{max}, D_{50}

References

Alipour, A. and Mokhtarian, M., 2021. RSM-based model to estimate the V-cut drill and blast pattern specific charge in rock tunneling. International Journal of Geomechanics, 21(11), p. 06021030.

Allen, M. and Worsey, P., 2015. Burn cut pull optimization through varying relief hole depths. Proceedings of the 41st Annual Conference on Explosives and Blasting Technique, New Orleans, LA, USA, February 1–4.

Bakhtavar, E., Abdollahisharif, J. and Ahmadi, M., 2017. Reduction of the undesirable bench-blasting consequences with emphasis on ground vibration using a developed multi-objective stochastic programming. International Journal of Mining, Reclamation and Environment, 31(5), pp. 333–345.

Berta, G., 1994. Blasting-induced vibration in tunnelling. Tunnelling and Underground Space Technology, 9(2), pp. 175–187.

Cardu, M., Godio, A., Oggeri, C. and Seccatore, J., 2021. The influence of rock mass fracturing on splitting and contour blasts. Geomechanics and Geoengineering, pp. 1–12.

Cardu, M., Saltarin, S., Todaro, C. and Deangeli, C. 2021. Precision rock excavation: Beyond controlled blasting and line drilling. Mining, 1(2), pp. 192–210. https://doi.org/10.3390/mining1020013.

Cardu, M. and Seccatore, J., 2016. Quantifying the difficulty of tunnelling by drilling and blasting. Tunnelling and Underground Space Technology, 60, pp. 178–182.

Cardu, M., Seccatore, J., Vaudagna, A., Rezende, A., Galvão, F., Bettencourt, J. and Tomi, G.D., 2015. Evidences of the influence of the detonation sequence in rock fragmentation by blasting – Part I. Rem: Revista Escola de Minas, 68, pp. 337–342.

Desai, S.K. and Badal, R., 2000. New Austrian Tunnelling Method (NATM) tunnelling in basochhu hydroelectric project, Bhutan. Tunnelling Asia 2000: Proceedings, New Delhi, p. 473.

Dey, K. and Murthy, V.M.S.R., 2012. Prediction of blast-induced overbreak from uncontrolled burn-cut blasting in tunnels driven through medium rock class. Tunnelling and Underground Space Technology, 28, pp. 49–56.

Emsley, S., Olsson, O., Stenberg, L., Alheid, H.J. and Falls, S., 1997. ZEDEX-A study of damage and disturbance from tunnel excavation by blasting and tunnel boring. Technical Report 97–30, Swedish Nuclear Fuel and Waste Management Co., Stockholm, Sweden, 218 pp.

Engin, I.C., Maerz, N.H., Boyko, K.J. and Reals, R., 2020. Practical measurement of size distribution of blasted rocks using LiDAR scan data. Rock Mechanics and Rock Engineering, 53(10), pp. 4653–4671.

Fjaer, E., Holt, R.M. and Raaen, A.M., 1989. Rock mechanics and rock acoustics. ISRM International Symposium, OnePetro Org., Pau, France, August 30–September 2.

Fornaro, M., Mancini, R., Cardu, M. and De Antonis, L., 1989. Ein Allgemeingultiges Stuck-grossenverteilungsgesetz und seine Anwendung in der Sprengparametersauswahl Hinsichtlich der Optimisierung der Haufwerksbildung in einem Steinbruch. Berg-und Huttenmannische Monatshefte, A. 134(2), pp. 31–34.

Godio, A. and Seccatore, J., 2019. Measuring the reduction of the confinement along the evolution of a burn cut. Applied Science, 9, p. 5013. https://doi.org/10.3390/app9235013.

Grady, D.E. and Kipp, M.E., 1979. The micromechanics of impact fracture of rock. International Journal of Rock Mechanics and Mining Sciences & Geomechanics Abstracts, 16(5), pp. 293–302. Pergamon.

Hagan, T.N., 1984. Means of increasing advance rates and reducing overall costs in drill and-blast tunnelling. Australian Tunnelling Conference, 5th, Sydney (No. 84/8).

Hino, K., 1956. Fragmentation of rock through blasting and shock wave theory of blasting. The 1st US Symposium on Rock Mechanics (USRMS), ARMA-56-0191, Golden, CO, USA, April 23–25.

Hino, K., 1959. Theory and Practice of Blasting. Nippon Kayaku Co., Ltd, Japan, 197 pp.

Hu, Y., Lu, W., Chen, M., Yan, P. and Yang, J., 2014. Comparison of blast-induced damage between presplit and smooth blasting of high rock slope. Rock Mechanics and Rock Engineering, 47(4), pp. 1307–1320.

Hunter, G.C., McDermott, C., Miles, N.J., Singh, A. and Scoble, M.J., 1990. A review of image analysis techniques for measuring blast fragmentation. Mining Science and Technology, 11(1), pp. 19–36.

Ibarra, J.A., Maerz, N.H. and Franklin, J.A., 1996. Overbreak and underbreak in underground openings part 2: Causes and implications. Geotechnical & Geological Engineering, 14(4), pp. 325–340.

Johansen, J. and Mathiesen, C.F., 2000. Modern Trends in Tunnelling and Blast Design. A.A. Balkema, Rotterdam, NL.

Kahraman, S., 2002. The effects of fracture roughness on P-wave velocity. Engineering Geology, 63(3–4), pp. 347–350.

Katsabanis, P.D., Tawadrous, A., Braun, C. and Kennedy, C., 2006. Timing effects on the fragmentation of small scale blocks of granodiorite. Fragblast, 10(1–2), pp. 83–93.

Kawalec, W., Krol, R., Suchorab, N. and Szymanski, M., 2019. The analysis and assessment of grain size distribution on the example of a chosen granite mine. IOP Conference Series: Earth and Environmental Science, IOP Publishing, Bristol, UK, November, vol. 362(1), p. 012113.

Kim, D., Noh, S., Lee, S., Park, B. and Jeon, S., 2007. Development of a new center-cut method: SAV-cut (stage advance V-cut). Romancov & Zlamal (Eds.). In Underground Space-the 4th Dimension of Metropolises, Three Volume Set+ CD-ROM, CRC Press, Taylor & Francis Group, Boca Raton, FL, USA, pp. 521–528.

Konya, A., Konya, C.J. and Worsey, P., 2017. Modern V-cut design. Engineering and Mining Journal, 218(9), pp. 40–42.

Konya, C.J. and Walter, E.J., 1991. Rock blasting and overbreak control (No. FHWA-HI-92-001; NHI-13211). United States. Federal Highway Administration.

Lee, C.I., Jong, Y.H., Jeon, S., Choi, Y.K. and Kim, H.S., 2005. The computerized design program for tunnel blasting. Proceedings of the 31th Annual Conference on Explosives & Blasting Technology, Orlando, USA, vol. 1, pp. 159–168.

Lou, X., Wang, B., Wu, E., Sun, M., Zhou, P. and Wang, Z., 2020. Theoretical and numerical research on V-Cut parameters and auxiliary cuthole criterion in tunnelling. Advances in Materials Science and Engineering, 2020, Article ID 8568153, pp. 1–13.

Lu, W., Chen, M., Geng, X., Shu, D. and Zhou, C., 2012. A study of excavation sequence and contour blasting method for underground powerhouses of hydropower stations. Tunnelling and Underground Space Technology, 29, pp. 31–39.

Ma, J., Li, X., Wang, J., Tao, Z., Zuo, T., Li, Q. and Zhang, X., 2021. Experimental study on vibration reduction technology of hole-by-hole presplitting blasting. Geofluids, 2021, Article ID 5403969, pp. 1–10.

Maerz, N.H., Ibarra, J.A. and Franklin, J.A., 1996. Overbreak and underbreak in underground openings Part 1: Measurement using the light sectioning method and digital image processing. Geotechnical & Geological Engineering, 14(4), pp. 307–323.

Maerz, N.H., Palangio, T.C. and Franklin, J.A., 1996. WipFrag image based granulometry system. Proceedings of the FRAGBLAST 5 Workshop on Measurement of Blast Fragmentation, AA Balkema, Montreal, Quebec, Canada, pp. 91–99.

Mancini, R. and Cardu, M., 2001. Scavi in roccia-Gli esplosivi. Hevelius Ed., Benevento, Italy, p. 153.

Mancini, R., Gaj, F. and Cardu, M., 1998. Atlas of Blasting Rounds for Tunnel Driving. Politeko, Torino, Italy, 337 pp.

Mandal, S.K. and Singh, M.M., 2009. Evaluating extent and causes of overbreak in tunnels. Tunnelling and Underground Space Technology, 24(1), pp. 22–36.

Mandal, S.K., Singh, M.M. and Dasgupta, S., 2008. Theoretical concept to understand plan and design smooth blasting pattern. Geotechnical and Geological Engineering, 26(4), pp. 399–416.

Martino, J.B. and Chandler, N.A., 2004. Excavation-induced damage studies at the underground research laboratory. International Journal of Rock Mechanics and Mining Sciences, 41(8), pp. 1413–1426.

Nefis, M. and Talhi, K., 2016. A model study to measure fragmentation by blasting. Mining Science, 23, pp. 91–104.

Paillet, F.L. and White, J.E., 1982. Acoustic modes of propagation in the borehole and their relationship to rock properties. Geophysics, 47(8), pp. 1215–1228.

Raina, A.K., Chakraborty, A.K., Ramulu, M. and Jethwa, J.L., 2000. Rock mass damage from underground blasting, a literature review, and lab-and full scale tests to estimate crack depth by ultrasonic method. Fragblast, 4(2), pp. 103–125.

Ramulu, M., Chakraborty, A.K. and Sitharam, T.G., 2009. Damage assessment of basaltic rock mass due to repeated blasting in a railway tunnelling project – A case study. Tunnelling and Underground Space Technology, 24(2), pp. 208–221.

Rehman, H., Ali, W., Naji, A.M., Kim, J.J., Abdullah, R.A. and Yoo, H.K., 2018. Review of rock-mass rating and tunneling quality index systems for tunnel design: Development, refinement, application and limitation. Applied Sciences, 8(8), p. 1250.

Rubin, A.M. and Ahrens, T.J., 1991. Dynamic tensile-failure-induced velocity deficits in rock. Geophysical Research Letters, 18(2), pp. 219–222.

Sanchidrián, J.A. and Singh, A.K., 2012. Measurement and analysis of blast fragmentation. The 10th International Symposium on Rock Fragmentation by Blasting, India, New Delhi, pp. 24–25.

Seccatore, J. and Cardu, M. 2018. On the influence of breaking angles in the pull efficiency of tunnelling by drill and blast. Proceedings of the XIII Jornada de Voladura, Viña del Mar, Chile, October 17–19; ASIEX – Chilean Association of Explosives Engineers, Santiago, Chile, pp. 76–80.

Seccatore, J., Cardu, M. and Bettencourt, J., 2015. The music of blasting. Sustainable Industrial Processing Summit, 4, Antalya, Turkey.

Shehu, S.A., Yusuf, K.O. and Hashim, M.H.M., 2020. Comparative study of WipFrag image analysis and Kuz-Ram empirical model in granite aggregate quarry and their application for blast fragmentation rating. Geomechanics and Geoengineering, pp. 1–9.

Silva, J., Worsey, T. and Lusk, B., 2019. Practical assessment of rock damage due to blasting. International Journal of Mining Science and Technology, 29(3), pp. 379–385.

Singh, S.P. and Lamond, R., 1993. Investigation of blast damage and underground stability. 12th Conference on Ground Control in Mining, Sudbury, Ontario, pp. 366–372.

Singh, S.P. and Xavier, P., 2005. Causes, impact and control of overbreak in underground excavations. Tunnelling and Underground Space Technology, 20(1), pp. 63–71.

Swift, R.P., Schatz, J.F. and Durham, W.B., 1979. Effect of simultaneous and sequential detonation on explosive-induced fracture: Proc 19th US symposium on rock mechanics, Stateline, Nevada, 1–3 May 1978, V1, p. 235–242. Publ. Reno: University of Nevada, 1978. In International Journal of Rock Mechanics and Mining Sciences & Geomechanics Abstracts, 16(2), April, pp. 43–44. Pergamon Press.

Swithenbank, J.J.M.D.S.D.G.C., Beer, J.M., Taylor, D.S., Abbot, D. and McCreath, G.C., 1976. A laser diagnostic technique for the measurement of droplet and particle size distribution. 14th Aerospace Sciences Meeting, p. 69.

Van Aswegen, H. and Cunningham, C., 1986. The estimation of fragmentation in blast muckpiles by means of standard photographs. Journal of the Southern African Institute of Mining and Metallurgy, 86(12), pp. 469–474.

Worsey, P.N., Farmer, I.W. and Matheson, G.D., 1981. The mechanics of pre-splitting in discontinuous rock. The 22nd US Symposium on Rock Mechanics (USRMS), ARMA81, Cambridge, MA, USA, June 29–July 2.

Xinkuan, Z.O.U., Jichun, Z.H.A.N.G., Qiang, P.A.N. and Wei, W.A.N.G., 2018. Vibration reduction effect of stepped V-cut blasting. Journal of Southwest Jiaotong University, 53(3).

Yang, L., Yang, A., Chen, S., Fang, S., Huang, C. and Xie, H., 2021. Model experimental study on the effects of in situ stresses on pre-splitting blasting damage and strain development. International Journal of Rock Mechanics and Mining Sciences, 138, p. 104587.

Yang, X., Hu, C., He, M., Wang, H., Zhou, Y., Liu, X., Zhen, E. and Ma, X., 2019. Study on presplitting blasting the roof strata of adjacent roadway to control roadway deformation. Shock and Vibration, 2019, Article ID 3174898, pp. 1–16.

Zare, S. and Bruland, A., 2006. Comparison of tunnel blast design models. Tunnelling and Underground Space Technology, 21(5), pp. 533–541.

Zhang, X., Pak, R.Y., Gao, Y., Liu, C., Zhang, C., Yang, J. and He, M., 2020. Field experiment on directional roof presplitting for pressure relief of retained roadways. International Journal of Rock Mechanics and Mining Sciences, 134, p. 104436.

Chapter 6

Blasting design

6.1 Introduction

The relevant parameters for a blast design can be synthesized as follows:

- geometry (shape and size of the charges and the volume assigned to them, position and extension of the free surfaces, position of the charges with respect to the volume to be blasted) – generally, the geometry is considered the most important factor;
- amount and type of explosive;
- delay time of detonators.

The blast must be designed carefully, and all details must be checked with respect to the wanted effect and the containment of unwanted effects within relatively strict limits.

At least since the 1500s, explosives have been exploited not only for their historical use (firearms) but also for excavation and demolition purposes. In the latter case, it was not possible to reach such mathematically sophisticated, precise, and elegant solutions as in the former, where the problem and the data were defined much more roughly: rifles and bullets, even of different military armies, are less different from each other than two blast holes of the same blast.

Both fields (weapons and blasting works) have witnessed evolution, as in almost all branches of technology, through a process that can be defined as "Darwinian". The habits and precautions that led to good results were replicated more frequently than the others until they have become common practices. Of course, this dominance persists as long as no change in the external environment occurs that alters the scenario, such as the affirmation of a different technique, the change in some restrictive conditions, the introduction of a new technology/innovation on the market, or some requirements of the result.

Therefore, going back to the specific case under examination, the rules for the calculation of blast holes and blasts should be understood above all as standards for adapting a solution that has given good results (e.g., the shooting plan used by a competitor and/or a neighbor with success) under slightly different conditions compared to the site under examination (e.g., different rocks, explosives, and bench height).

It is then very important to have available, in addition to the rules for the calculation, a collection of solutions that have been proven valid. For this reason, some examples and application schemes coming from real cases in history have been added to this chapter and they can be found at the end of each paragraph of this section.

DOI: 10.1201/9781003241973-6

6.2   Characteristics of the rocks in term of blastability

Rocks are usually described with nouns and adjectives but, when it is necessary to evaluate how to technically operate on them, nouns and adjectives must correspond to (approximate) intervals of quantitative indicators of some of their relevant properties.

In a blast design problem, it is important to have three types of indicators available:

- strength indicators of the rocks;
- indicators of integrity, or mechanical continuity, of the rock mass (presence of joints and/or discontinuities);
- indicators of the preferential position, if any, of the discontinuity surfaces of the rock mass with respect to the excavation face – this information is very important to carry out stability assessments and for any need to orient the geometry of the face differently to the extent possible, for example trying to take advantage of the weak planes when organizing the blast, or to increase or reduce the height of the bench, to take special precautions to regularize it, and so on.

As for the strength, the Protodyakonov (Boky, 1967) classification can still be used; it was initially developed for mining but also adopted in civil works, and it is shown in Table 6.1.

This classification divides the rocks into eight classes of increasing easiness of excavation (the use of the explosive is related to the first 6 classes) and is based on the compressive strength (the Protodyakonov index, reported in Table 6.1, corresponds to 1/10 of the compressive strength in MPa, determined in the laboratory with canonic tests or alternatively with expeditious tests).

The first class, "not-classified rocks" or Class 0, includes exceedingly strong rocks, which are difficult to both drill and break down.

As for the integrity, it is currently expressed by the fracture density $[m^{-1}]$ (Boadu, 1997), that is, the number of fractures that occur on average over the length of 1 m, and the classification currently used is shown in Table 6.2.

As for the detection of the preferential position of the discontinuities, the most concise, smart, easy, and user-friendly indication for an evaluation, or decision, is obtained by reporting in spherical projection the pole of the main family of discontinuities, the pole of the face, and the direction of progression of the excavation, as clarified by the example shown in Figure 6.1.

6.3   Sizing formulas

Considering the topic of the charges' calculation, the formulas and procedures proposed below have been chosen based on the "simplicity" of the theories from which they arise and their convenience of use. Obviously, many other theories and formulas equally valid and acceptable exist and can be used (Chung and Katsabanis, 2000; Segui and Higgins, 2002; Hudaverdi et al., 2012; Bowa, 2015; Lu et al., 2016).)

In most formulas, the charge is unknown, but, in many cases, it is necessary to consider the geometry as unknown, since the charge is bounded by various kinds of restrictions.

An important caution is that no formula can be used "blindly": evaluating whether the solution is practically applicable is always necessary.

Consider an example: suppose that the calculations have shown that a certain pattern can be performed with holes of 30 mm in diameter, 20 m in length, with a burden of 60 cm, each charged with 15 cartridges of 28 mm in diameter and length of 500 mm, triggered with in-the-hole delayed detonators. Before proceeding with an executive drawing, it is necessary to check if those holes are feasible, if it makes sense that they are drilled 60 cm from the wall, if the

Table 6.1 Classification of rocks for the purpose of breaking. According to its original definition, the Protodyakonov strength number, f, is the instantaneous compressive strength c [kg/cm²] of the rock, divided by 100: $f = c/100$. Every rock can be described by its Protodyakonov strength number, which is obtainable by laboratory tests.

Strength classes of rocks	Excavation technique	Description of rock	Protodyakonov index	Swell factor
1	2	3	4	5
Not-classified Class 0	Explosives	Exceedingly strong quartzites and porphyrites, gabbro-diorites, gabbro-diabase, olivine basalt, andesites, hornfels, diabases, the strongest diorites	20–25	2.2
		Exceedingly strong, fine-grained granites	17–18	2.2
		Flints, compact exceedingly strong quartzitic sandstones, silicified limestones	15–16	2.2
Class I	Explosives	Medium-grained granites, compact quartzitic sandstone, quartzites, diabase, strong gneiss, porphyrites, strong trachytes, syenites	12–14	2.2
		Fine-grained quartzitic sandstones, compact very strong limestones, exceedingly strong marbles	10–11	2.2
Class II	Explosives	Strong calciferous conglomerates, strong quartzitic sandstones, strong pyrite, dolomite, and hard limestone	8–9	2.0
		Coarse-grained serpentine, granite and syenite	7	2.0
Class III	Explosives	Strong argfillites and siltstones, siliceous clay slates, siderites, magnesites, serpentine with talc, dense limestones	6	2.0
		Granites, gneisses, syenites and other highly mineralized or weathered igneous rocks	5	2.0
		Marly limestones, dolomites and clayey micaceous shale	4–5	2.0

Strength classes of rocks	Excavation technique	Description of rock	Protodyakonov index	Swell factor
1	2	3	4	5
Class IV	Explosives and pneumatic/ hydraulic hammer	Clayey and carboniferous shales of medium strength, dense marl, weak siliceous shale, weak limestone, and dolomite	– –	– –
		Anthracite, strong coal, weak sandstone or conglomerate, medium-strong siltstone, and argillite	2	1.8
Class V	Explosives and pneumatic/ hydraulic hammer	Weak clay shale, very weak limestone and dolomite, medium-strong coal, strong lignite	1.5–2.0	1.4
		Dense carbonate clays, dense chalk, medium-strong marl, gypsum, strong rock salt	1.5	1.8

(Continued)

Table 6.1 (Continued)

Strength classes of rocks	Excavation technique	Description of rock	Protodyakonov index	Swell factor
1	2	3	4	5
Class VI	Pneumatic/ hydraulic hammer	Soft coal, hard loess, soft marl, lignite, carbonate clays, diatomite, soft rock salt, porous gypsum	– –	– –
		Dense clays, glacial loam, grease clays, heavy loam containing more than 10% of pebbles or gravel, craie (marl, etc.)	– –	1.4
		Demolition rubble	1.0–1.5	1.8
Class VII	Hand-held tools such as pneumatic picks	Light clays, loams, sandy loams, loess, coarse gravel, gravel, ballast	0.9	1.8
		Sand, loose sand, topsoil	0.6	– –
		Calcareous tuff and other friable rocks	0.4	– –

Source: Boky (1967)

Table 6.2 Rock classification based on fracture density

Fracture density (m^{-1})	Designation
<1	Massive
1–2	Slightly fractured
2–4	Moderately fractured
4–6	Fractured
6–10	Highly fractured
10–20	Extremely fractured
>20	Milonitized

Source: Modified from Billaux et al. (1989)

Note: In general, explosives are used limitedly to the rocks of the first five classes (from massive to highly fractured rocks).

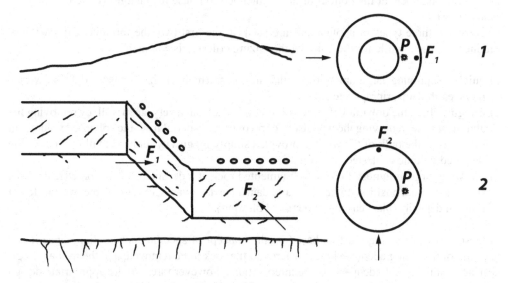

Figure 6.1 Perspective view (left) of two possible orientations of an excavation face (F_1 and F_2) in rock with iso-oriented fractures. Representation of the two situations using the spherical projection (right) of the fracture poles (points P) and the poles of the faces (points F_1 and F_2). The arrows indicate the directions of advancement of the faces in the two cases; case 1 is the most hazardous for those who work from the yard, and case 2 is the safest.

introduction of those cartridges and the related triggering systems are feasible, etc. (the example is intentionally meaningless).

6.3.1   Calculation of spherical charges

6.3.1.1   Formula of the Military Engineering Corps

The formula (Barnes, 1947; Powell, 1948) links the charge (Q, in kg) to the burden, that is, the minimum distance of the center of gravity of the charge from the free surface (V, m).

The formula is used for calculating large heading blasts. The volume shot down by a given charge depends on the local topography; therefore, the specific consumption is not used for the design: obviously, it is calculated, after calculating the charge or the burden, depending on how the problem was posed, based on the local topography data, and can be varied by changing the position of the charges.

The formula is expressed as:

$$Q = abV^3$$

where V is the burden and a and b are empirical coefficients (a depends on the explosive; b depends on the rock class); their values (usually adopted) are given in Table 6.3. The above-reported formula is also called the "monomial formula of Military Engineering Corps".

The large heading blasts are prepared by carrying out tunnels with a small cross section, and by digging chambers at the bottom of them, intended to house the charges (hundreds or thousands of kg).

Currently, three types of applications are used: in the first two, the formula of the Military Engineers is used, while for the third another sizing criterion is adopted:

1) quickly supplying large amounts of rock in coarse (metric) size for bank defense works, piers, earth dams, ripraps, etc.;
2) bringing about the downfall of large volumes of rock from a very steep wall, by removing the abutment base, removing the wall itself, or recovering material from the collapsed volume. In these cases, the bulk of the work is provided simply by gravity: only a small fraction of the removed volume is blasted;
3) throwing large volumes of coarsely fragmented rock at a decametric – hectometric distance from its original position, to rapidly build dams, barriers, yards, or remove the overburden of mineral deposits (blast casting, directional blasting).

In these cases, the sizing must be based on the concept of specific consumption, since it is a question of providing adequate kinetic energy to the rock to be removed, and the military engineering formula is not adequate. These interventions, however rare, require appropriate design criteria (for any further information, the use of specific literature is recommended, including Nedriga et al., 1983; Adhikari and Gupta, 1989; Ninahua and Yongji, 1994; Yue et al., 2013; Yang et al., 2019).

Table 6.3 Values of empirical coefficients *a* and *b* for different explosives and rock types.

Explosive	a coefficient	Rock class	b coefficient
Straight gelatine	0.16	Soft ground (VII, VI)	1.75
Ordinary dynamite	0.24	Tuff (V)	2
TNT	0.3	Soft rock (IV)	2.5
Nitrate-based explosives	0.56	Average rock (III; II)	3.3
		Hard rock (I, 0)	4.25
		Fractured rock (*)	5.6

Source: The data are mediated from various sources (Wright et al., 1953; Olson et al., 1970; Olson and Fletcher, 1971; Adhikari et al., 1999); the numbers in brackets refer to Protodyakonov's classification.

(*): with open fractures that can lead to relief.

A basic scheme of these operations and examples are shown in Figures 6.2 and 6.3.

The powder factor is generally in the order of a few hundred g/m³ in the first category mentioned, in the second it usually falls below 100 (obviously, considering as "blasted rock" also

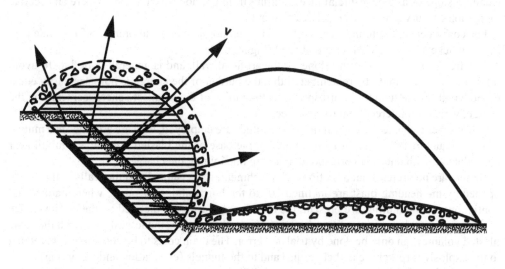

Figure 6.2 Principle diagram of a directional blasting operation. A system of large coplanar charges blasted simultaneously (schematized in the section by the thick blackened band) and simulating a single large tabular charge crushing and ejecting the rock outside and producing a long heap with a controlled shape and size (Chernigovskii, 1986).

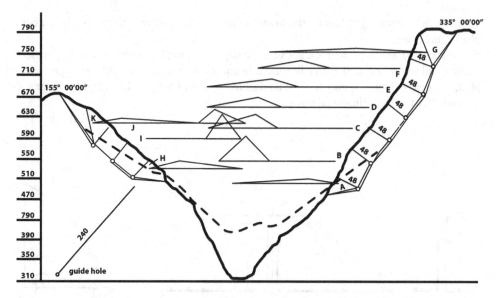

Figure 6.3 Example of application of directional blasting to realize a river barrier. On the two sides of the valley, two systems of heading blasts have been prepared according to the principle described in Figure 6.2 which, exploding simultaneously, eject the rock into the groove, obstructing it. The triangles represent the sections that would ideally assume the heaps produced by the single charges, if they would detonate separately; the dimensions are in meters (Xu, 1991).

the volume that falls by gravity, and once the abutment base has been removed), in the third it rises to thousands of g/m³.

Regarding the first application, which is the most common, the tunnels start from the wall of a natural slope or a steep artificial face resulting from previous excavations, where an accurate topographical survey was previously carried out.

For convenience, the tunnels are horizontal when possible and can branch off into side passages (brackets) if more charges are needed (Figure 6.4).

The free surface is that of the slope, or the artificial wall, and is generally not flat. To avoid frequent errors, it should be remembered that the burden of the charge is not necessarily equal to the length of the tunnel: it corresponds to the radius of the smallest sphere, centered on the charge, which can intercept the free surface.

When several charges aligned at the same altitude are employed, they are arranged at a mutual distance equal to 1–2 V, and it is assumed that the base of the blasted volume is a mixtilinear quadrilateral, with one side corresponding to the line joining the charges.

If there are no preferential directions of detachment, it is assumed that the walls of the cavity opened by the heading blast are inclined 70° to the horizontal. Obviously, where natural and continuous discontinuities are present, the detachment surface is dictated by them. The choice of the most suitable position of the charges (the one that ensures the lowest cost, with the same blasted volume) can only be done by trial and error. The main costs to be considered are related to the explosives (purchase and charging) and to the tunnels (excavation and stemming).

The position of the charge must however satisfy various conditions, in addition to that of the favorable geometry of the volume to be blasted:

- It has to be placed in a rock that is not affected by faults or discontinuities, to avoid the risk of relief.
- To avoid the stemming "blown", the tunnel must have at least one abrupt deviation before placing the charge.
- The vertical thickness of the rock above the charge must not be too high with respect to the burden, unless the desired effect is rockslide and there are no favorably arranged natural planes of discontinuity. It may also be advisable, in case of very high walls, to arrange more than one blast on superimposed planes.

(Westwater, 1957)

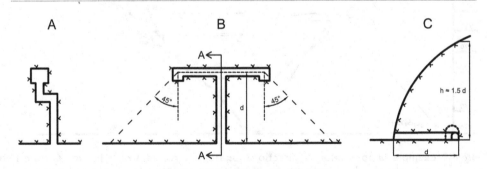

Figure 6.4 Typical layouts of large heading blasts (Seguiti, 1969); A: a single blast-hole (plan); B and C: a blast-hole with two brackets (plan views and vertical section along the access tunnel).

The volume of blasted rock, corresponding to that on-site increased by the bulking factor, is arranged in a heap with a 30°–40° rest angle; of course, it is necessary to verify in advance that there is space enough to accommodate it.

Instead of the monomial formula of the Military Engineering Corps, some designers prefer other criteria, also empirical, of the binomial type: the charge is calculated as the sum of two terms, one proportional to the third power of the burden, and the other to the second power, with coefficients other than those shown above. Furthermore, when the heading blasts are systematically used in a project, it is sometimes preferred to calculate the charges by first evaluating the volume to be blasted, then multiplying it by a specific consumption obtained from the analysis of the first blast.

Figure 6.5 describes an amazing application of the heading blasts, used for leveling the Pao Tai Mountain (China, 1992) due to the construction of an airport. The project was a directional blasting operation, aimed to throw the biggest possible volume of fragmented rock into the sea to gain space. Up to date, it is probably the largest civil blasting project (141 charges, 12,000 t of explosive, 33 micro-delays with a total duration of the heading blast of 3.8 s, 11 Mm^3 of granitic rock removed, and a specific consumption close to 1 kg/m^3 (usual range of directional blasting works).

6.3.1.2   Calculation of spherical charges according to the crater theory

The crater theory (Livingston, 1956) is a systematization of the intuitive concepts on which the formula of the military engineers is based, that is, the concept of the range of action of the charge. According to this theory, the effect of a spherical charge of a given explosive, detonating in a given rock at a given distance (burden) from an indefinite free wall, is described by two indexes (which can be determined by blast tests) characterizing the explosive-rock pair:

- the ratio between the volume of the crater produced by the charge and the mass of the charge;
- the ratio between the depth at which the center of the charge is disposed of, and the cube root of the mass of the charge.

(Hino, 1956; Starfield, 1966)

The first index is referred to as the *scaled volume*, the second is referred as the *scaled depth*, and the graphical representation that takes both indexes into account (with the depth scaled on the horizontal axis and the volume scaled on the vertical axis) is known as a *crater curve* (Katsabanis and Liu, 1998; Ye, 2008; Kamarudin et al., 2012; Wang et al., 2018).

The crater curve has an asymmetrical shape, starting from the origin, with a gradually increasing trend up to a maximum, which is reached for an optimal depth value (the maximum crater volume for a given charge) and a rapid descent beyond the maximum until it returns to zero when the scaled depth is too large to produce observable external effects. Beyond this point, of course, the scaled volume remains zero (Cardu, 1990).

Since the crater curve for a given explosive-rock pair may be known from blast tests or from literature, the scaled depth must be established so that it falls in the growing branch of the crater curve, just before the maximum: it is not advisable to use the value corresponding to the maximum, because a small, accidental deviation from the ideally expected behavior of the curve could bring the actual behavior of the charge on the descending branch of the curve, with a strong risk of failure.

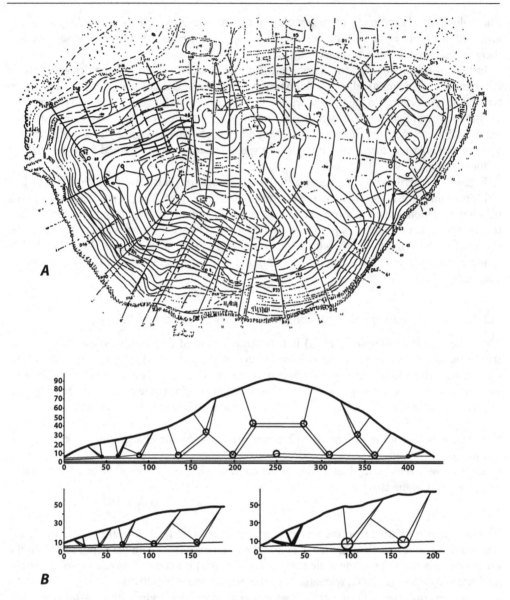

Figure 6.5 Leveling Mount Pao Tai (China, 1992) for the construction of an airport, with a single blast of 141 large heading blasts. This is a "directional blasting" operation, aimed
at projecting most of the volume of rock into the sea to gain space (total charge of
12,000 tons of explosives, organized according to 33 delays with a total duration of
3.8 s, to remove a volume of 11 million m³ of granite). The specific consumption,
around 1 kg/m³, is in the usual range of directional blasting works. In A, the plan
of the intervention, with the position of the access tunnels to the chambers. In
B, some sections are shown with an indication of the volumes of influence of the
charges. The dimensions are in meters (Sheng and Gu, 1993).

On the other hand, if the blast hole is only intended to obtain the effect of inducing fractures in
the rock while leaving it in place, without expelling fragments, the scaled depth must be beyond
the point where the curve descends to zero.

Figure 6.6 shows examples of crater curves (Clark, 1987), relating to medium-strength rocks (classes III–IV of Table 6.1).

An example of how to use this method is developed below, for the case of a spherical charge consisting of a cylindrical cartridge with a length equal to six times the diameter.

The known data are

- the crater curve (Figure 6.6);
- the hole diameter Φ (generally, in the range of 165–170 mm);
- the diameter/length ratio of the charge Φ/l (it has been found that the most effective value is approximately 1/6).

With reference to Figure 6.6, it was decided to refer to the point x, since, for certainty of effect, it is better to work below the peak value. By drawing the parallels to the vertical and horizontal axes (dashed lines), two points, a and c, are intercepted, which correspond, respectively, to:

$$a = \frac{D}{\sqrt[3]{W}};$$

$$c = \frac{V_k}{W}$$

from which the following parameters are obtained:

D = depth at which the center of gravity of the charge must be positioned from the surface $(D = a\sqrt[3]{W}$ [ft], [m]);
V_k = volume blasted by one blast-hole $(V_k = W \times c$ [ft³], [m³]);
W = mass of the charge (known value).

When the length of the charge is equal to six times the diameter, it can be written as:

$$W = \gamma_{espl} \frac{3\pi}{2} \Phi^3$$

It is then possible to find the necessary diameter and the necessary hole length L $(L = D + 3\Phi)$.

The characteristic curve, therefore, allows to solve the problem of sizing a blast-hole; however, it is formulated:

1) Given W, find D: $D = a\sqrt[3]{W}$.

2) Given D, find W: $W = \left(\dfrac{D}{a}\right)^3$.

3) Given W, find V_k: $V_k = C \cdot W$.

4) Given V_k, find W: $W = \dfrac{V_k}{C}$.

5) Given Φ, find W: $W = \gamma_{espl} \cdot \dfrac{3\pi}{2} \cdot \Phi^3$.

6) Given Φ, find D: $D = a \cdot \Phi \cdot \sqrt[3]{\dfrac{3\pi \cdot \gamma_{espl}}{2}}$.

7) Given Φ, find V_k: $V_k = C \cdot \gamma_{espl} \cdot \dfrac{6\pi}{4} \cdot \Phi^3$.

8) Given W, find Φ: $\Phi = \sqrt[3]{\dfrac{2W}{3\pi \cdot \gamma_{espl}}}$.

9) Given V_k, find Φ: $\Phi = \sqrt[3]{\dfrac{2V_k}{3\pi \cdot c \cdot \gamma_{espl}}}$.

10) Given D, find Φ: $\Phi = \dfrac{D}{a \cdot \sqrt[3]{\gamma_{espl}} \cdot \dfrac{3\pi}{2}}$.

The relationships 6 and 10 establish, for the same rock and explosive, a direct proportionality between the diameter of the hole and the depth at which the center of gravity of the charge must be placed to obtain the wanted effect, which is the analog of the empirical criterion followed to assign to the burden of cylindrical blast holes a value proportional to the diameter of the hole.

6.3.1.3   Underground crater blasting

The use of the crater theory in open pit excavations is unusual. Conversely, it can be employed in underground as an upside-down crater, at the crown of the excavation: in this case, the explosive yield is greater than in upwards craters (Lang et al., 1977; Lang, 1982), as shown in Figure 6.7.

The volume of rock between the face of the excavation and the supporting arch that originated the crown is in very precarious stability conditions: it is subjected to tensile stress longitudinally, (due to its weight) and transversely (the rock after the blasting tends progressively to bend as the fractures and joints in the rock tend to open). It can be stated that in the upside-down crater, the action of the explosive is considerably facilitated: following the blast, in fact, not only the crater due to the charge used is obtained (true crater in Figure 6.7), but it is also possible to detach the rock from the fractured area above (ruptured zone in Figure 6.7) that would not fall if the crater had been facing upwards. Sometimes, also occurs the collapse of the rock of the stressed area that generates because of the above tensile stresses, resulting in the formation of a cavity whose height can be up to three times greater than the depth of the crater (stressed zone in Figure 6.7). However, this may lead to less predictability of the blasting effect, while improving, as mentioned, the action of the individual charges.

For a production blast, a distribution of interfering craters can be used (Figure 6.8): thanks to the fractures parallel to the face (which in this case is the crown), the craters do not remain as isolated cones: in fact, the entire volume of rock below the level where the charges are positioned is blasted, consequently the next blast finds a theoretically flat surface. This type of blast is called with the acronym VCR (vertical crater retreat or vertical retreat stoping). This method is based on the cooperation of spherical charges: the construction site is therefore organized upwards by horizontal slices.

The stoping starts from below and advances upwards (Figure 6.9); when necessary, the broken ore can partly remain in the stope to support the walls: the ore is recovered at the bottom or underneath the stope, through the draw point system, and is generally hauled using LHD loaders.

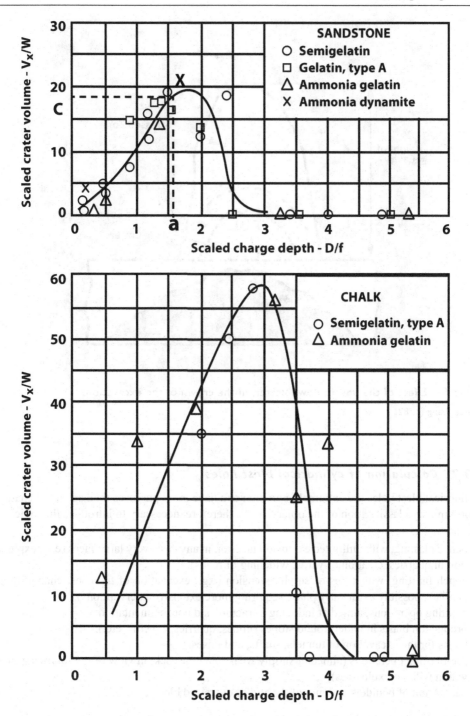

Figure 6.6 Examples of "scaled volume/scaled depth" curves (Clark, 1987). Above: from sandstone tests; below: from tests on weak limestone. A possible operating point to
be used as a means of proportioning the blast hole is indicated on the sandstone
curve. It is important to note that both, the scaled volume V_k/W and the scaled
depth D/r reported, are not expressed in metric units, but in ft³/lb and inches,
respectively.

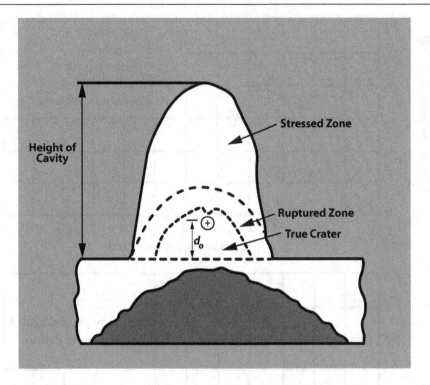

Figure 6.7 Effect of the upside-down crater, in the crown of the excavation.
Source: Lang (1982)

6.3.2 *Calculation of cylindrical blast holes*

Cylindrical blast holes are extremely common and have more purposes than spherical charges. A preliminary classification of the objectives is, therefore, necessary to introduce the topic:

- bench blasting, with unlimited extension faces or, in any case, very large faces (e.g., exploitation of quarries, creation of yards, widening of roads);
- bench blasting, with faces of small extension (e.g., excavation of trenches, channels, and similars, possibly as an early phase requiring more extensive interventions);
- splitting operations, aimed at inducing a fracture and not fragmentation;
- production blasts in underground stopes (mines, quarries, caverns, etc.);
- blasts for the excavation of tunnels, shafts, and raises;
- pre-blasting operations (aimed at simply weakening the rock, in view of the following excavation without explosives);
- demolition of boulders and monoliths and/or oversized blocks.

It should also be noted that the last two operations can often be performed with spherical charges.

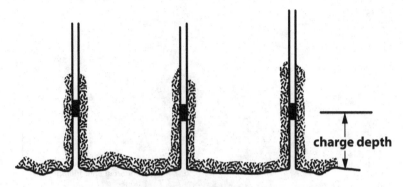

I.   STOPE BACK PRIOR TO BLAST

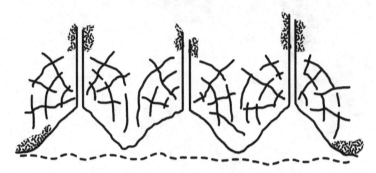

II.   INITIALS CRATERS AND FRACTURE ZONES

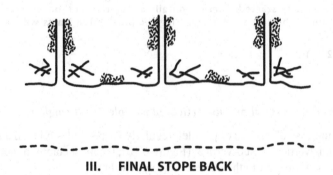

III.   FINAL STOPE BACK

Figure 6.8 Cooperation between contiguous craters V.C.R. blasting. The initial configuration (I) evolves, upon blasting, to configuration II, but the radial cracks and those caused by the shock wave reflection make the residual volumes between the craters unstable under their own weight, and part of the rock above the level of the charges. This rock collapses immediately, leaving a theoretically flat crown, as in configuration III

Source: Crackel et al. (1981)

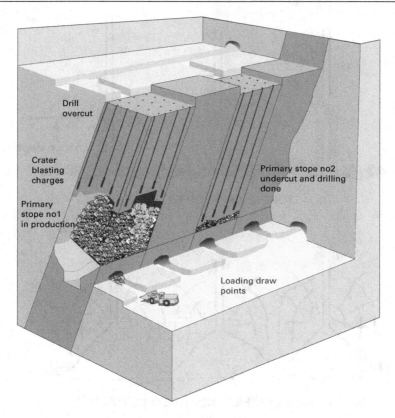

Figure 6.9 Typical scheme of application of the V.C.R. in a vein or in a straightened layer of strong orebody The scheme exemplifies an underground stope where production blats have already started. Further details about the method of exploitation and organization of the stope before starting the production blasts will be provided in Chapter 7.

Source: Atlas Copco (2007)

6.3.2.1   Bench blasting: vertical or sub-vertical blast holes on a single row

This is a very common case of a series of parallel, regularly spaced holes lying in a plane parallel to the face, which is vertical or sub-vertical (Figure 6.10); the most common inclinations are between 70° and 90° with respect to the horizontal (Figure 6.11).

It is necessary, first, to spend some words on the acknowledged relationships, which are not to be considered design rules, but simply ranges of some geometric parameters that should be respected, not only for reasons related to the use of explosives: the ratio r_1 of the burden V (distance between the axes of the holes and the free surface) to the spacing E (distance between the axes of the holes) should preferably be between 0.8 and 1 in isotropic rock. A value of 1 (square pattern) is often preferred to facilitate the drilling design in multi-row blasts. This will be discussed later.

Essentially geometric considerations on the "radius of action" would recommend, as mentioned, even lower values of r_1, up to = 0.7, but such a low ratio is not practical, as portions of

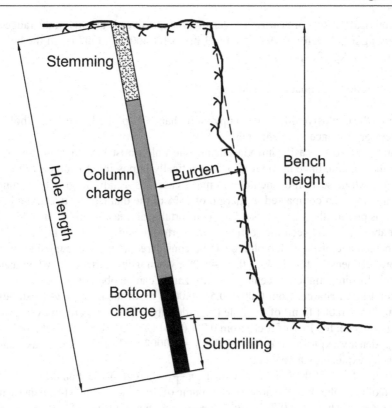

Figure 6.10 Typical blast hole in surface mining.

non-blasted rock could remain between two holes. On the other hand, since the correct behavior of a blast depends, no matter the theory used for the calculation, on the ratio of the charge to the burden (increasing the burden necessarily involves increasing the charge), it is advisable to keep this ratio as low as possible: assuming that the suitable charge requires a certain burden V, since the volume blasted by a hole of unit length is the product between the spacing and burden (and reducing the spacing E means reducing the volume assigned to the blast-hole), increasing the ratio beyond what is necessary is equivalent to the unnecessary consumption of more explosives. However, the anisotropy of the rock can require modifying the optimal ratio: for instance, if the rock is vertically stratified, with the layers orthogonal to the face, the spacing should be reduced to prevent layers from remaining intact between the holes in the row (Berta, 1990); if, on the other hand, the layers lie parallel to the face, the spacing can be increased:

- The volume of rock V_1 assigned to 1 m of blast-hole, which is obviously equal to $V \times E$, can be written as: $V_1 = V^2/r_1$, and it usually varies between V_2 and $V_2/0.8$.
- Whatever the sizing formula is chosen for the design, there is a relationship between the amount of explosive to be used and the volume of rock to be blasted: it is called specific consumption, or powder factor (P.F.), and varies according to the rock characteristics, the explosive and other factors, but generally, for blasts of this type, it is in the range of 0.2–0.4 kg/m^3.

- The theoretical density, or "cartridge" density of explosives is usually in the range of 900 kg/m³–1,400 kg/m³; therefore, $1/900 = 0.0011$ m³ to $1/1,400 = 0.0007$ m³ of the hole would be necessary for 1 kg of explosive.

However, the following must be considered:

- In general, the explosive does not occupy more than 70% of the length of the hole, since it.is also necessary to place the stemming.
- The cartridges, to be easily introduced into the hole, must have a diameter considerably smaller than it, and, once introduced, it is practically impossible to compact them until they occupy the whole section of the hole. In many cases, especially in gravity loading, the cartridges are not even compacted. A margin of 15% of the diameter between the hole and the cartridge is practically unavoidable, for a comfortable and fast loading, and corresponds to a ratio of about 0.72 between the section of the cartridge and that of the hole; this means that only a percentage of the hole volume can be considered actually occupied by the charge at its theoretical density, that is $70 \times 0.72 \cong 50\%$, when using cartridges (when using bulk or pneumatic loading, higher percentages of utilization can be obtained).
- Then, for 1 kg of charge, from $0.0011/0.5 = 0.0022$ m³ of the hole (low-density explosives) to $0.0007/0.5 = 0.00142$ m³ of the hole (high-density explosives) are on average necessary; as a consequence, for 1 m³ of rock, from $0.2 \times 0.00142 = 0.000284$ m³ of the hole (minimum P.F. high-density explosive) to $0.4 \times 0.0022 = 0.00088$ m³ of the hole (maximum P.F. and low-density explosive) are needed.
- The volume of 1 m of the hole is, obviously, $\pi \frac{\Phi^2}{4}$, where Φ is the diameter.
- The ratio of the volume of the hole to the volume of the rock assigned to a generic unit length of the hole, varies between $0.785 \times \Phi^2/V^2$ (pattern with $r_1 = 1$) and $0.8 \times 0.785\ \Phi^2/V^2$ (pattern with $r_1 = 0.8$).

By equating this ratio to the requirement previously calculated, the following results, as extreme cases, are obtained:

- $0.8 \times 0.785\ \Phi^2/V^2 = 0.00088$, that is, $\Phi^2/V^2 = 0.0014$ and $\Phi/V = 0.037$, pattern with $r_1 = 0.8$, low-density explosive, maximum P.F.;
- $0.785 \times \Phi^2/V^2 = 0.000284$, that is, $\Phi^2/V^2 = 0.00036$ and $\Phi/V = 0.019$, squared pattern ($r_1 = 1$), high-density explosive and minimum P.F.
- The ratio r_2 of the burden to the diameter (V/Φ), commonly referred to in calculations (e.g., for the choice of an adequate drilling diameter) varies accordingly, in the range of $1/0.037 = 27$ to $1/0.019 = 53$ (but more commonly it is contained in the range 30–40).
- The ratio r_2 is also subject to other limitations: a too-high value can induce the blast hole to violently eject the stemming and rock fragments of the uncharged part of the hole, creating a crater around the mouth of the hole; a too-low value, even locally, can give rise, for the whole charge or for a portion of it, to an excessive concentration of charge between the blast-hole and the free surface, and this can result in fly rocks that can travel over a great distance. This risk is real when r_2 falls (locally) below 20 and this places limits on other ratios that will be examined later. In the case examined (bench blasting with sub-vertical holes on a single row), the burden practically differs very little from the horizontal distance between the axis of the hole and the face; r_2 can be evaluated without serious errors by using the latter distance instead of V, which is more convenient in the design of the blast.

- The ratio r_3 of the length of the stemming (i.e., the distance between the mouth of the hole and the last cartridge introduced) to the burden (B/V) must be contained within certain limits, dictated by various considerations: to exert an appreciable containment of the reaction products, the stemming length should not be less than 20 Φ; since the minimum burden is 30 Φ, the minimum stemming length should be at least 0.66 V. This limit is also suggested by the need not to reduce the minimum distance between the charge and the free surface below 20 Φ, as mentioned above. However, it is not advisable to follow the minimum value: frequently, the edge of the bench is significantly lower than the burden of the hole as a whole, due to "back-break" effects, which are more sensitive in the upper part of the face, and it is also frequent in the case where very high benches must be divided into horizontal slices, and that the upper portion of the underlying bench is physically the tread of the overlying bench; therefore, the rock is damaged for a certain depth by the action of the previous blast. In most cases, it is therefore prescribed that the length of the stemming is not less than 0.75 V. For precaution, stemming is often assigned a length equal to V. It is not convenient to exceed the length of the stemming, for two reasons: the first, obvious, is that by increasing its length, the length available for the charge decreases; the second is that increased stemming can result in the permanence of dangerous rock protrusions in the upper part of the bench, or in large blocks not fragmented but simply detached and loosened by the explosive. Hence, r_3 is generally kept in the range of 0.75–1.0.
- The ratio r_4 between the length of the hole (or, practically, the bench height) and the burden (L/V) must also be contained in a given range, for the following reasons: since the length of the stemming corresponds to at least 75% of the burden, in order to contain the necessary amount of charge, the length of the hole must be considerably greater: for at least 70% to be used, this length must be, for obvious geometric reasons, at least equal to 2.5 V, by assuming $r_3 = 0.75$, and at least 3.3 V, if $r_3 = 1$ is assumed. In practice, values of r_4 less than 3 are rarely adopted; as for the upper limit of r_4, apparently, it does not exist, except for the one given by the drilling rig: however, this is not the case. The face of the bench, which represents the free surface, is not an ideal plane, but an irregular surface that approximates a plane for an easier calculation of volumes, distances, and charges. It represents the residual wall from the previous blast and contains recesses and protrusions; even the axis of the hole is never perfectly straight and exactly inclined with respect to the vertical, as assumed by the project. Consequently, the value of the actual distance between the axis of the hole and the wall, that is, the value of the burden, becomes increasingly uncertain. The uncertainty is minimal at the mouth of the hole, which is based on measurements; operating with normal accuracy, and with a face that is not too irregular, it can be admitted that it increases by 5 cm for every meter of length: for example, if V corresponds to the design value in the initial section of the hole, at a depth of 10 m, it can still have the correct value, but it can also be 50 cm shorter or 50 cm longer (in heterogeneous or fractured rocks the deviation is often greater, due to the greater difficulty in drilling accuracy and to the irregularity of the bench). Serious problems can occur if the real burden differs from the designed one by less than 1/3 (danger of throwing stones) or more than 1/3 (bad detachment at the base of the bench, which requires further dangerous interventions). To operate correctly, therefore, the third part of V must be less than 5% the length of the hole. The ratio r_4 must not exceed 0.33/0.05 = 6.6, rounded to 6 as prudence. Therefore, the usual range of r_4 is between 3 and 6.
- Generally, bench blasting is aimed at removing a given volume of rock at a given level, leaving a flat and regular surface at that level. To obtain this result it is only necessary, theoretically, to make the holes reach the agreed plane, but the explosion rarely cuts the rock cleanly at the base according to a horizontal plane. More often, it shears it according to an inclined surface dictated

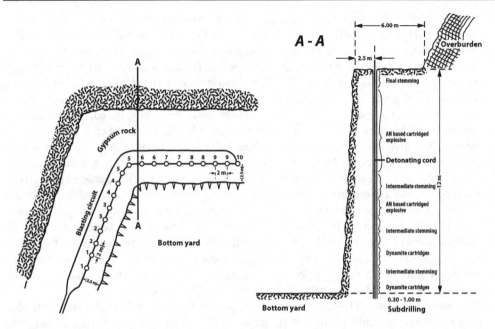

Figure 6.11 Example of a blast with vertical parallel holes on a single row, in a gypsum stone quarry (Spannagel, 1969). Right: plan view and blasting circuit. Left: vertical section of a blast hole. Charges: ammongelit (dynamite) + donarit (nitrate explosive), divided into four decks with intermediate stemming (Zwischen besatz) and final stemming made by drilling debris (Bohrmehel besatz). Detonating cord (Sprengschnur) along the hole.

Table 6.4 Ranges commonly adopted for single-row bench blasting.

Geometric size (A)	A/Burden	A/Diameter
Burden – V	-	30–40
Spacing – E	1–1.25	30–50
Diameter – Φ	0.033–0.025	-
Stemming length – B	1–0.75	22–40
Hole length – L	3–6	90–240
Subdrilling – S	0–0.3	0–12

by the line of less strength, leaving a salience whose height is greater than the burden. Consequently, in order to ensure compliance with the agreed plan it is necessary to extend the holes below this plan for a length S (subdrilling) which can vary from 0 to 1/3 V; there is, therefore, a supplementary ratio $r_s = S/V$, with a range between 0 and 0.33.

The ranges of the geometric relationships analyzed are shown in Table 6.4.

The holes can be vertical or slightly inclined; vertical drilling is easier to perform, but inclined drilling has various advantages: a better detachment at the base of the bench, a more stable face (in a well-done job, the slope of the face corresponds to that of the holes), a smaller deviation of the effective bench edge with respect to its theoretical position. A too-high inclination, however, can cause charging difficulties. The usual range of inclination is between 0° and 25° with respect to the vertical; the most frequently adopted value is 20°.

The procedures proposed and used to obtain the values to be assigned to the charges are numerous. Among them, the simplest have been selected, which can be classified as follows:

- static sizing criterion;
- methods based on the concept of specific consumption.

6.3.2.2   Static sizing criterion

The objective is to obtain separated slabs in the muck pile, with thicknesses dictated by the natural spacing of the joints or weak planes. "Burned" rock should be minimized with respect to the entire blasted volume, yet high explosives cannot be avoided due to the strength of the rock. Rock types range from metamorphic to volcanic and are usually hard and abrasive; for this reason, quarrying is usually by blasting (very cautiously) rather than by mechanical cutting.

The holes are drilled orthogonal to the natural joint planes. The ratio of the burden to the spacing of holes is kept very high, which implicates a lower powder factor and percentage of "burned" rock (the rock close to the holes, badly fractured by the charges) with respect to the whole blasted volume.

The most common features of the blasts are hole diameters of 50–90 mm, hole spacing of 1–2 m, bench height of 15–25 m, hole length up to 10 m, P.F. 50–100 g/m³; explosives used are straight dynamites and water gels; a line of detonating cord is placed in the hole, along the whole length, to warrant complete detonation; ignition is from the collar, with electric instant or ms caps; stemming occupies 40–60% of the hole length (Cardu et al., 1995; Cardu and Castelli, 1996).

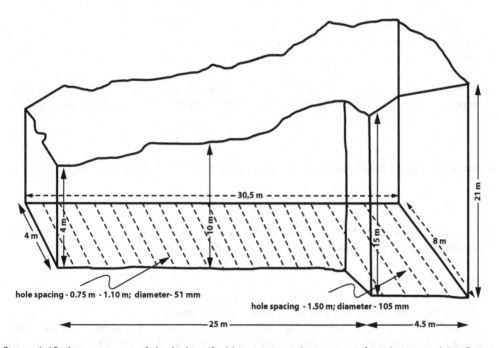

Figure 6.12 Arrangement of the holes of a blast in a porphyry quarry for obtaining slabs. Since, in this case, the holes are drilled at the base of the volume to be removed, S₁ is the horizontal surface, while S₂ and S₃ are respectively the rear and the side surfaces (in the example, three free surfaces are available before the blast).

The example shown in Figure 6.12 refers to a blast of horizontal holes for the production of porphyry slabs.

The calculation, for a blast of vertical holes, proceeds as follows: from the geometrical and strength data, the quasi-static borehole pressure needed to obtain the rock failure according to the intended surface is obtained.

The surface that is subjected to tensile stresses (S_1) is the one that contains the charges, whereas the other surfaces (S_2, horizontal, and S_3, lateral) are subjected to shear; each of them is multiplied by its strength to the stress considered (the tensile strength for S_1, the shear strength under load equivalent to the lithostatic pressure for S_2, and the simple shear strength for S_3).

It should be noted that the rocks are often mechanically anisotropic and that those for the production of natural slabs are almost always anisotropic: therefore, the strength values for each type of stress depend on the orientation of the surface considered.

The three contributions to the force thus calculated are concordant by direction and are added together, obtaining the total resisting force F_r.

The explosive intended to be adopted is chosen, which is characterized by a certain explosion pressure P_e (the quasi-static pressure is considered, at the ordinary loading density). Useful values for the calculation are in the following:

- dynamites – 1,400 MPa;
- medium-strength explosives – 1,000 MPa;
- ANFO – 900 MPa.

As for the in-the-hole density, the following values can be assumed:

- dynamites – 1,200 kg/m³;
- medium-strength explosives – 1,000 kg/m³;
- ANFO – 800 kg/m³.

By F_r by the explosion pressure, the active surface S_a is obtained, that is, the surface on which the explosive must act to obtain the desired active force, capable of balancing the resisting force:

$$S_a = F_r/P_e$$

S_a corresponds to the product of the holes' diameter Φ by the total charged length L_c:

$$L_c = Sa/\Phi$$

From L_c, the total amount Q of explosive is obtained, multiplying the volume of the hole corresponding to this length by the density of the explosive in the hole:

$$Q = 0.785 \times \Phi^2 \cdot \delta$$

where δ is the in-the-hole density (loading density) of the charge.

The total length of holes L_t is obtained by adding to L_c what is necessary to contain the stemming. This can be done by setting a suitable utilization coefficient for the holes, for example, ¾, (1/4 intended for stemming and ¾ for explosives), and, in this case, it can be written as:

$$L_t = L_c \times 4/3$$

Obviously, the increase in length to accommodate the stemming can also be established with other criteria.

Finally, by dividing S_t by L_f, the spacing between the holes is obtained.

The same procedure can also be applied to the design of blasts with horizontal holes: in this case, the surface subjected to tensile strength is the horizontal one, to which the weight of the volume to be blasted and the vertical surfaces (subjected to shear strength) must be added.

The static sizing criterion does not consider the concepts of optimal spacing between the holes, nor that of Powder Factor: ten holes with a diameter of 100 mm spaced by 2 m would be equivalent to 20 holes with a diameter of 50 mm spaced by 1 m, but the former contains an amount of charge 4 times higher (according to this, the minimum Powder Factor would be obtained with infinite holes of null diameter!).

However, it is necessary to avoid solutions that involve a too large spacing between the holes, to eliminate unevenly detached surfaces: E should not exceed $\cong 20\ \Phi$; more often, it is contained in the range between 10 and 15 Φ.

The sizing principle is like the one used for splitting blasts.

As for the strength values to be assumed, for isotropic rocks, if test data are not available, the following may be recommended:

Tensile strength: $\cong 0.1$ compressive strength;
Simple shear strength: $\cong 0.5$ compressive strength;
Coefficient of friction: $\cong 0.5$.

6.3.2.3   Methods based on the concept of specific consumption

These methods are the simplest and most practical, which assume that, for a certain explosive/rock pair and in similar geometric conditions, the amount of explosive to be used is proportional to the volume to be blasted. Very simple calculation formulas derive from this hypothesis, provided that the goal is simply to make a certain volume of rock removable.

The procedures become more complicated when the expected result is shown in terms of blasting a certain volume of rock, to be converted into a granular material characterized by a predetermined (more or less strictly) particle size distribution, that is, when the goal is to use the explosive, at the same time, as a medium of disconnection and as a crusher having an adjustable outlet size.

As it can be easily understood, the particle size distribution of the material depends on at least four factors:

- the explosive-rock pair;
- the blasting pattern;
- the original density of fractures' distribution;
- the powder factor.

Hence, in this method, the powder factor is no longer a constant feature of the explosive-rock pair. The calculations regarding the first, simpler case, are reduced to two, with some variations in detail: the monomial formula and the binomial formula.

6.3.2.3.1   CALCULATION ACCORDING TO THE MONOMIAL FORMULA

When the benches are low and the holes have a small length, so that each blast hole affects limited volumes (a few m³), they are charged with a continuous charge of a single explosive, simply calculated by multiplying the blasted volume by a specific consumption suitable for the explosive-rock pair:

$$Q = \text{P.F.} \cdot H \cdot E \cdot V$$

where

- Q = charge, kg;
- P.F. = powder factor adopted, kg/m³;
- H = bench height, m;
- E = spacing between the holes, m;
- V = burden, m (if the holes are vertical; otherwise, when they are sub-vertical, a small correction must be applied, often negligible, due to the non-coincidence of the burden with the horizontal distance between the hole and the face).

H is generally known, while E and V, linked together by the chosen value of the pattern ratio, must be calculated. The diameter can be selected from a limited number of values, corresponding to the bits available with the drilling machine adopted or can be pre-set.

Since H and the inclination of the holes are known, if a certain subdrilling is necessary, the length of the hole is easily calculated. An appropriate stemming length is deducted from it so that the length to be charged is obtained.

The volume of the charged hole is calculated by choosing the diameter and multiplying the section of the hole by the length to be loaded; the charge to be used is obtained by dividing this volume by the density of the explosive in the hole. Once the charge is known, E and V can be calculated.

The problem can, however, be posed differently: the charge could be imposed, for example, due to the limits dictated by vibrations: in this case, the volume is obtained using the same formula and, once H is known, V and E are deduced. On the basis of V, the stemming length is chosen and the chargeable length is calculated.

The volume of the charge is calculated from its weight and in-the-hole density; therefore, once the chargeable length of the hole is known, its diameter is deduced: obviously, it must correspond to one of the diameters available, otherwise the assumptions about the blasting pattern and the stemming length must be modified within reasonable limits (obviously this applies if the data of the problem are not seriously incompatible with a reasonable solution).

6.3.2.3.2   CALCULATION ACCORDING TO THE BINOMIAL FORMULA

It is known experimentally (Langefors and Kihlström, 1967) that the specific consumption necessary for a blast hole, with a length not much greater than the burden, is considerably greater than that required by a blast hole that must blast the rock from a stretch of a high wall placed far above the bottom of the face. It is said that in the first case the blast hole works with a fixation at the bottom, whereas in the second case, it works with a free burden at the bottom (Figure 6.13). This is intuitive since in the first case a large part of the rock must be broken by shear stresses, while in the second it yields only under tensile stresses (and the shear strength of the rocks is much greater than its tensile strength). Following up, the specific consumption necessary for a blast hole, with the same rock and explosive, decreases as the ratio of the height of the bench to the burden increases. In fact, as this ratio increases, often considered a slenderness ratio, the percentage incidence of the volume of rock to be blasted with a fixation at the bottom decreases.

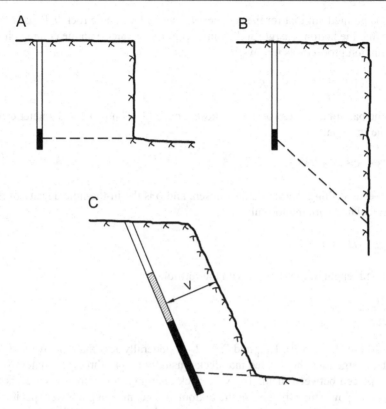

Figure 6.13 Blasting in a vertical bench, with a fixation at the bottom (A) and free burden at the bottom (B); bench blasting with inclined holes (C). The dashed lines represent the trace of the release plane.

To transform these qualitative observations into a practical calculation criterion, Langefors and Kihlström (1967) proposed computed the calculation of the charge with a formula that uses two distinct specific consumption values: one higher for the lower part of the bench and the other lower for the remaining part. This formula, therefore, calculates the charge as the sum of two portions and for this reason, it is called binomial. This approach is commonly adopted, with some adjustments depending on the specific case, in the design of bench blasting with heights greater than 5 m, frequent in the exploitation of quarries and large civil excavations. As for the portions of the bench to which the highest specific consumption (bottom charge – Q_1) and the lowest (column charge – Q_2) are assigned, the first is given a height equal to the burden, and the second, obviously, the counterpart. The formula can then be written as:

$$Q = Q_1 + Q_2 = \text{P.F.}_1 \cdot V^2 \cdot E + \text{P.F.}_2 \cdot (H - V) \cdot V \cdot E$$

where P.F._1 e P.F._2 are the two powder factors adopted. They can be known from blast tests (performed in the two conditions of fixed and free burden at the bottom) or from the analysis of similar cases.

It has to be pointed out that for the same explosive and the same rock, $P.F._1$ is $\cong 2 \div 2.5 \, P.F._2$. This means that the bottom charge, at the same length, is greater than the column charge.

The calculation proceeds as follows:

$$Q_1 = V^2 \cdot E \cdot P.F._1 = V^3 \cdot P.F._1 \cdot R$$

where R is the pattern ratio, appropriately chosen; once Q_1 is known, the diameter can be calculated with the formula:

$$Q_1 = 0.785 \cdot \Phi^2 \, (V+S) \cdot \delta$$

where S is the subdrilling, appropriately chosen, and δ is the in-the-hole density of the charge. Q_2 is then calculated with the formula:

$$Q_2 = V \cdot E \cdot (H-V) \cdot P.F._2$$

and the charged length of the hole L_c, with the formula:

$$L_c = \frac{Q_2}{0.785 \cdot \Phi^2} \cdot \delta$$

Due to the difference between $P.F._1$ and $P.F._2$, L_c is generally less than the available length. In this case, the charge must be divided into decks (usually two or three), and interlayers of inert material are placed between them. To avoid deck charging, a common rule refers to using a denser and more powerful explosive at the bottom charge, meaning a lower specific consumption than that used for the remaining part. This need is generally satisfied by using dynamite for Q_1 and ANFO for Q_2, provided that the holes are dry and have a diameter suitable to ensure the good detonation of ANFO. This dynamite-ANFO scheme is commonly adopted. Of course, other types of mixed charging are possible, even with three or more types of explosives in the same hole.

6.4   Most frequent blast geometries

6.4.1   Blasts with vertical or sub-vertical holes, on multiple rows

In bench blasting with several rows, when possible, a potential solution to be used refers to blast holes that can be sized with the monomial or the binomial formula, as in the blasts on a single row, assuming the distance between the rows as a burden. This solution is implemented to obtain higher productivity (or, in other words, to intervene faster). In fact, due to the delays between the rows, when the blast holes of the rows following the first detonate, the volume of rock assigned to the previous row is already broken, even if not yet removed.

Compared to the case of blasts on a single row, the powder factor is increased by about 30%, as the face is not completely free. Moreover, due to the bulking factor, in a blast over many rows, the height of the muck pile can even exceed the height of the face: therefore, in this case, there is also a lifting effect of the blasted rock, and its energy cost may require a further increase in terms of powder factor. The pattern is usually kept the same for all rows and is usually squared. An example is provided in Figure 6.14.

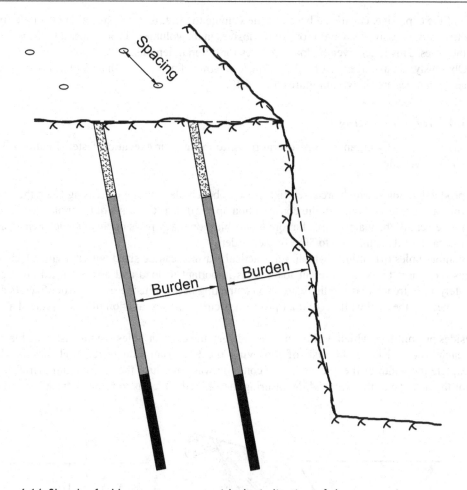

Figure 6.14 Sketch of a blast on two rows with the indication of the geometric parameters.

6.4.2   Bench blasting with horizontal holes or with parallel fan blasts

In general, these schemes are used when vertical or sub-vertical holes cannot be applied, for example, due to the inability to access the upper portion of the bench. These schemes involve various drawbacks: they force the drillers to work under the bench, which imposes a much more accurate scaling which is, at the same time, also more difficult due to the low regularity of the face left by the blast-holes of the previous blast; moreover, the holes are more difficult to drill, and also both charging and stemming are more complex.

Blasts with horizontal blast holes are sized like vertical blasts, by usually adopting the monomial formula.

As for the fan blasts, a bottom charge and a column charge must be distinguished in each blast hole. The blasts are drilled so that the toes of the holes are arranged in an approximately squared pattern, while the mouths of the holes are close to each other: therefore, the volume of rock that the unit length of the hole must blast decreases from the bottom to the mouth. Furthermore, the holes have different lengths. The bottom charge is calculated as for vertical holes; the remaining

part of the explosive, calculated based on the volume to blast, is distributed along the holes trying to avoid excessive concentrations of explosives in the volumes of rock closest to the mouths
of the holes. This is achieved by loading holes for different lengths.

Obviously, the average use of holes is lower in terms of percentage than the blasts of vertical
holes. Examples are shown in Figures 6.15–6.17.

6.4.3 Trench blasting

Blasts are generally organized on several rows, to make their execution faster (Figures 6.18–
6.21); they include

- production holes, which are sized like those of bench blasting but increasing (with the same
  rock and explosive) the specific consumption up to 50%, as the walls of the trench hinder the
  movement of the material much more than a blast on many rows on an extended bench, and
  increasing subdrilling (up to 50% of the burden);
- contour holes (trimming), which, for practical reasons, can be sized with the same diameter
  as the others but with different loads: it is important to load and arrange them appropriately, as in trench blasting there are two contrasting needs of using high Powder Factors and
  respecting the final walls. A similar problem occurs in the excavation of tunnels and shafts.

Besides presplitting (which will be discussed later), the contour holes are the last to be blasted
in each row and have the task of removing the two triangular-based rock prisms that
complete the width of the excavation. A common way to reduce the charge, and therefore to
limit the damage to the walls while ensuring the desired effect, is to replace these holes with

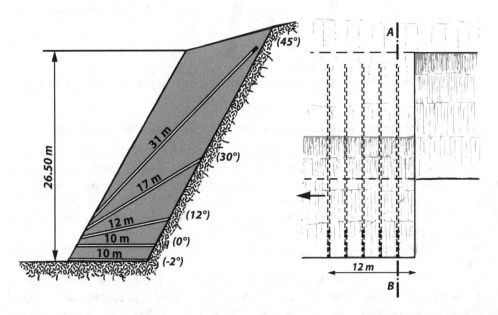

Figure 6.15 Large fan-shaped blast to produce limestone in a quarry (Seguiti, 1969). Drilling diameter: 100 mm; explosives used: ammonium nitrate and TNT explosive;
charges/hole: F_1 and F_2 = 12 kg; F_3 = 15 kg; F_4 = 23 kg; F_5 = 42 kg; total charge:
520 kg.

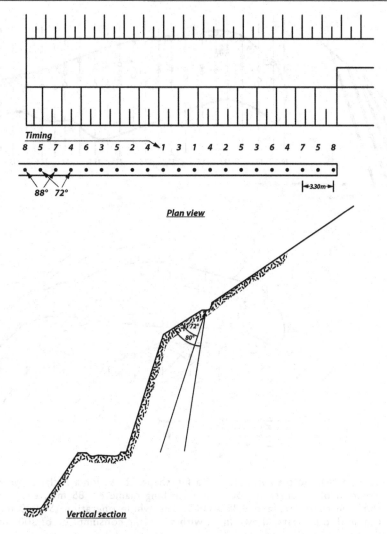

Figure 6.16 Blast for the enlargement of a roadway, with holes of different inclinations (plan view and vertical section). This scheme was adopted at higher altitudes, where the drillers could only access a narrow track, and the "toe spacing" at the height of the road would have been too much. Note the trigger order that, in each pair of holes, always involves the explosion first of the hole inclined by 72°, which has the smallest burden

Source: Büsch (1977)

pairs of holes initiated simultaneously, aligned on the contour line, one on the row and the other on the half of the row-to-row burden; obviously, their charge is much lower and can consist in decks or in decoupling. Stemming must be limited (even only equal to 50% of the burden) to have a precise cut of the edge of the wall, without the risk of back-break. For the excavation of very narrow trenches (2–3 m), schemes are used such as those shown in Figure 6.21, and the charging is usually performed with the monomial formula.

To obtain a better profile, two rows of pre-splitting holes can be added, loaded with tubular charges.

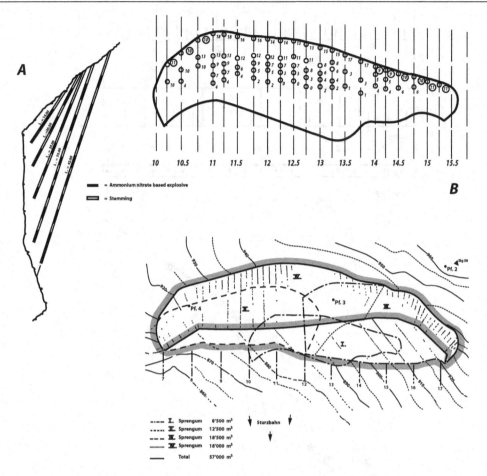

Figure 6.17 Section (A) and plan view (B) of a fan-shaped blast for a reclamation operation (removal of an unstable rock face). Drilling diameter: 85 mm; explosives used: slurry in cartridges and bulk ANFO. The whole operation (57,000 m³ of rock) required four blasts, shown in C, with a specific consumption of 300 g/m³.

Source: Modified from Rotzetter (1977)

6.5 Presplitting and splitting

Two cases must be distinguished:

- The fracture is an operation performed before an ordinary blast, where the volume isolated by the fracture must be fragmented, during the same intervention (adopting a correct delay scale) or subsequently.
- The fracture must isolate and slightly displace a volume of rock that must then be removed as a whole, intact.

The first case commonly refers to civil or mining works where the rock that remains in place must be protected from the eventuality of damage due to the blasts of the volume to be removed, and it consists of ordinary presplitting.

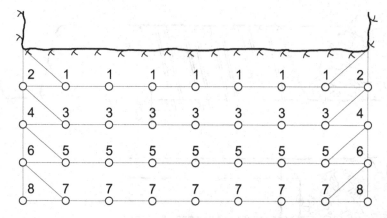

Figure 6.18 Example (plan view) of the arrangement of the holes for a large-width trench. The blast is on four rows of holes; the numbers refer to the detonation sequence: up to five holes can detonate simultaneously.

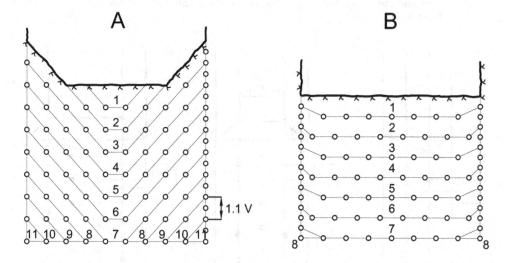

Figure 6.19 Examples of the arrangement of the holes for the excavation of large-width trenches, with thickening and reduction of the charge of the contour holes ("smoothing" or "smooth blasting") on one side only (A) or on both sides (B). Scheme A refers to a blast on ten rows, with inclined holes and spacing between the holes greater than the burden, which requires 11 delay numbers; the blast schematized in B is organized in seven rows and requires eight delay numbers. Of course, countless variations are possible.

The second case refers to the dimension stones sector (splitting), where the exploitation of intact blocks is applied for squaring and/or cutting into slabs, curbs, or other architectural items.

In both cases, the design of the charges applies directly to the principle of the static criterion, and the most important factors for obtaining the wanted result are the unit linear charge, the

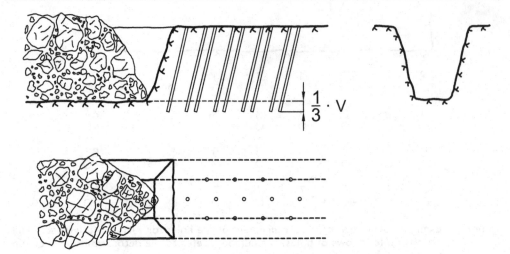

Figure 6.20 Scheme of a blast for the excavation of a narrow trench (e.g., for laying pipelines) in three perspectives. This is the classic pattern with staggered holes between row and row (1-2-1-2 . . . .). Sloped walls simply result from the overbreak of blast holes.

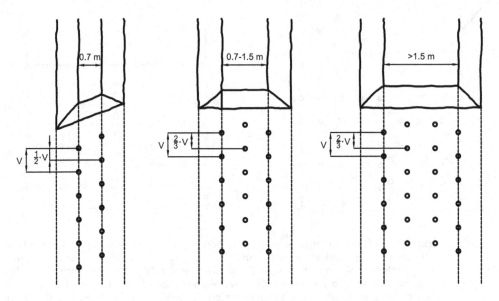

Figure 6.21 Other schemes (plan view) of blasts for the excavation of narrow trenches.

drilling diameter, and the relationship between the spacing and the diameter; the concepts of "burden" and "powder factor" are not useful.

The explosive requires a strong decoupling. The initiation of the blast holes is simultaneous or with a minimal delay. The drilling diameter is generally small, but in some cases, diameters can be greater than 100 mm: in such cases, the linear charge is not made up of tubular charges expressly designed for presplitting, but of ordinary cartridges equipped with detonating cord, or bulk explosives poured into pipes with a diameter much smaller than the hole, or of diluted explosives poured in bulk, etc. In typical pre-splitting operations carried out using tubular charges expressly manufactured for this purpose, manufacturers generally provide the data necessary for the use (diameter of the holes, spacing); a summary is contained in Table 6.5. In lack of such charges, for example in the case of pre-splittings made with ordinary cartridges and detonating cord (or similar solutions), the sizing can be done with empirical formulas, including the one obtained from the principle of the static criterion, having general validity (Del Greco et al., 1983):

$$\left(\frac{\Phi_c}{\Phi_f}\right)^2 \cdot \frac{L_c}{L} \cdot \frac{\delta_c}{1000} \cdot P_s \cdot \left(\frac{\Phi_f}{E - \Phi_f}\right) = T$$

where Φ_c is the diameter of the cartridges (m), Φ_f is the diameter of the holes (m), Φ_f is the density of the cartridges (kg/m³), P_s is the detonation pressure of the explosive (MPa), E is the spacing (m), L_c is the charged length (m), L is the hole length (m), and T is the tensile strength of the rock (MPa); in lack of data on it, the value of 20 MPa can be assumed to guarantee the wanted effect. As for the detonation pressure, different values for different explosives are provided, and in synthesis, the following can be assumed: 1,200 MPa for straight dynamites, 1,000 MPa for ordinary dynamites, 900 MPa for powdered nitrate explosives and for water gels, and 800 MPa for ANFO.

When the splitting for the exploitation of dimension stones is designed, the detonating cord (containing PETN) is generally used as a charge, and water or sand is used as stemming, or simply air, that is, not filling the empty space between the cord and the hole (in the case of weak rocks).

Table 6.5 Data for sizing the charges in presplitting operations

Hole diameter		Charge	Pipe charge		Hole spacing	Specific drilling	Cracked zone
Mm	in	kg/m dyn.	Ø (mm)	Length (mm)	m	drm/m²	m
32	1 ¼	0.13	22	380	0.45–0.7	2.22–1–43	0.4
32	1 ¼	0.21	25	1,140	0.45–0.7	2.22–1.43	0.4
38	1 ½	0.21	22	380	0.45–0.7	2.22–1.43	0.4
38	1 ½	0.21	25	1,140	0.45–0.7	2.22–1.43	0.4
51	2	0.38	32	1,000	0.5–0.8	2.00–1.25	0.5
51	2	0.47	32	380	0.5–0.8	2.00–1.25	0.6
64	2 ½	0.38	32	1,000	0.5–0.7	2.00–1.43	0.5
64	2 ½	0.47	32	380	0.5–0.7	2.00–1.43	0.6
64	2 ½	0.55	25	1,000	0.7–0.9	1.43–1.11	0.9
76	3	0.55	25	1,000	0.7–0.9	1.43–1.11	0.9
76	3	0.71	40	380	0.6–0.9	1.67–1.11	1.1
89	3 ½	0.90	32	1,000	0.7–1.0	1.43–1.00	1.7
89	3 ½	1.32	50	380	0.8–1.1	1.25–0.91	1.9
102	4	0.90	32	1,000	0.7–1.0	1.43–1.00	1.7
102	4	1.32	50	380	0.8–1.1	1.25–0.91	1.9

Source: Sandvik – Tamrock Corp. (1999)

The sizing formula, in this case, implies a set of simple, reasonable assumptions:

1) to break and move the unit volume of rock requires a certain amount of work;
2) the unit mass of the explosive can release a certain amount of energy;
3) the energy released is spent with a certain efficiency to perform the work quoted in (1).

Briefly, the blast is a machine fuelled by the explosive to perform rock breakage and displacement.

Any blasting design criteria based on the powder factor concept can therefore be labeled as "energetic". Alternatively, the system can be seen according to a different view: the explosive can be seen as a means to apply a force (a pressure by surface product) whose value must exceed the resisting force (a strength by resisting cross-section product) of a solid body that must be broken.

This concept is ineffective, as far as blast design is concerned, apart from one special case: the dimension stone production. Quarrymen engaged in this activity keep a record of powder factor in order to evaluate the unit cost of the production but decide the amount of explosive needed mostly on the basis of the split surface required to separate the intended rock volumes.

Any blast design criteria based on the breaking force concept can be labeled as "tensional" or "static", being practically like a structural stability check.

It is noteworthy to mention that the charge calculation formula cannot be as simple as $Q = c' * S$, with c' representing a "surface factor" in g/m^2 and S the split surface: 1 kg of explosive provides, in principle, a constant and known amount of energy, but can provide quite different amounts of breaking force, depending on the geometry of the holes and charges, hence c' is not a property of the explosive, rather of the pair hole and charges system. For a well-defined splitting practice (same rock, same drilling, same stemming, same charges), that simple formula should, in principle, hold.

Concerning the splitting method applied to isolate and slightly displace a volume of rock, experimental research was performed with the aim of finding practical, generic rules for blast design in granite quarries producing blocks (Mancini et al., 1992, 1994, 1995; Mancini and Cardu, 1995). The drilling (hole diameters 30–40 mm), hole spacing (20–30 mm), stemming (water), and charge (detonating cord, 10–24 g/m) are quite similar in most quarries, hence a specific "optimal surface powder factor" seemed easy to evaluate by simply working statistically with the data available on the results obtained from a great number of quarries, most of them using the detonating cord method since the 1960s. It was immediately apparent that things were not so simple: first, a controlled displacement of the block is required, and it is different from one case to another, hence the split surface is not the only parameter on which the charge depends; second, even disregarding the displacement requirement (which means, restricting the analysis to cases with comparable displacement) a minimal effective powder factor shows up. On a purely empirical basis, a polynomial charge design formula was proposed by Mancini et al. (1993) containing a constant term, a splitting surface-related term, and a displacement-related term:

$$P.F. = a + b \cdot \frac{S}{V} + c \cdot s$$

where $a = 10.5$, $b = 26.4$, $c = 28.7$, S (split surface) in m^2, V (volume) in m^3, and s (displacement) in m, a reasonable powder factor P.F. (g/m^3) for granite splitting operations. The term "a" represents a minimum effective specific charge and can be considered analogous to the idle running power of any machine. Reference is made to the volume simply because of a charging formula adhering to the powder factor principle was researched. The term "*b* * *(S/V)*" represents the share of the fracturing effect; the term "*c* * *s*" represents the share of the displacement effect.

Let us apply the formula to a typical average case, the separation of a 10 m wide × 20 long × 5 m high prism, 1,000 m³ weighing around 2,600 *t*, to be displaced by 0.10 m.

The total cut surface is 350 m², *s/V* is 0.35 m⁻¹, and the formula gives P.F.= 22.65 g/m³; hence, a total charge of 22,650 g of PETN, of which 2,874 g is required by displacement.

To move, merely by sliding, by 0.1 m the 2,600 *t*, with a supposed rock-to-rock friction coefficient of 1, requires around 2.6 MJ; actual demand is no doubt lower since a part of the travel is covered by flight, at a lower energetic cost. Being around 6.3 MJ/kg the specific explosion energy of PETN, propulsive efficiency is no more than 14%.

Polynomial formulas (the one that was worked, as well as others developed by other authors) are by themselves absurd: the underlying idea is that there are three effects wanted from the charge (to compensate losses, to split, and to displace) and the requirement of each one of them can be separately calculated and then added to obtain the total charge. But the three effects are, at best, successive effects of the same charge. The practical results simulate, in a way, the supposed behavior of the cord. It was decided not to abandon the polynomial formula concept, but rather to improve the fit by introducing a further parameter, rock strength, that was not considered at the beginning. Strong and weak granites exist, and average granite is a vague term.

According to the static approach, a blast hole charged at its full length should be considered upon explosion as a pressurized cylindrical container. If many parallel holes are prepared and exploded in a solid medium, the resultant force exerted orthogonally to the plane is simply given by the product of the length of the cumulated hole by the hole diameter and by the pressure (the first two terms of the product provide what can be named the "active surface" of the system).

By dividing this resultant surface by the cumulated area, in the holes' plane, of the inter holes' rock bridges, tensile stress is obtained; if failure of the bridges is desired, said tensile stress should exceed the tensile strength of the rock. Force can be increased either by reducing hole spacing or by increasing hole linear charge, in both cases increasing the specific surface charge. An improved formula is therefore expected by introducing the strength parameter in the empirical polynomial formula, in addition to the surface-to-volume ratio and displacement. The mean value of tensile strength (bending test) of commercial granites has been found to be around 13 MPa (Lundborg, 1967; Merriam et al., 1970; Sha et al., 2020). The range, in the quarries studied, was 6–15 MPa. The tensile strength can vary, even in the same quarry, from one point to another, and even from a part of a block to another. In any case, the average values for each quarry can be defined. Tensile strength data were collected, along with the blasting data, to perform the strength tests on representative samples of the production, when necessary. Working within the same general frame of the preceding statistical study, the most suitable expression of a function linking the powder factor to the parameters S, V, s, already defined, and T_0, the tensile strength (MPa), was found:

$$P.F. = 18.66 + T_0 \left(0.87 \frac{S}{V} + 2.19s \right)$$

It represents a fair improvement with respect to the simpler formula previously found. The new formula shows that the basic idea of the static calculation (to calculate the charge based on the theoretical breaking force and of the split surface) is an oversimplification: apparently, rocks do not resist splitting only by virtue of their tensile strength, but also the size of the block to be separated has a strong influence on this effect. The simple empirical rule of keeping a constant charge/split surface ratio retains practical usefulness only for the individual quarries, where the parameters involved (V, S, s, T_0) are not too much different from one blast to another. The same formula however shows how the correct value of the charge needed to separate and move by the desired distance one block of given size, shape, and strength, using the detonating cord in the way explained before, can be easily calculated using the data available to the quarryman.

As for the case of the pre-splitting, the detonating cord can also be used for splitting operations for different purposes (Figure 6.26). Figures 6.22–6.25 show typical examples of presplitting applications for trench blasting and for profiling the final walls of quarries at the end of the exploitation; furthermore, Figure 6.26 provides examples of splitting in granite quarries using the detonating cord/water technique.

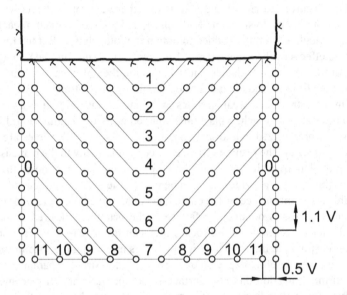

Figure 6.22 An example of the application of presplitting for profiling the walls of a wide trench; the pre-splitting rows are triggered at the time zero and cover the whole longitudinal development of the blast.

Figure 6.23 Example of application of presplitting with inclined holes for profiling a narrow trench. The production holes are arranged according to one of the schemes previously examined; the pre-splitting holes cover the whole longitudinal development of the blast, but not the whole excavation height (to avoid the risk of flashover). From left to right: plan, cross section, and longitudinal section.

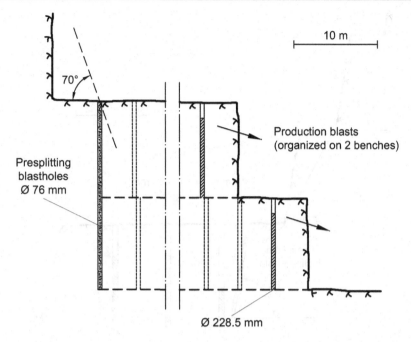

Figure 6.24 An example of the application of presplitting to provide the final profile of a quarry; it is implemented, in this case, long before proceeding with the production blasts, and covers the height of 2 benches at a time; given the considerable height (over 20 m), the diameter of the holes is large (159 mm).

6.6   Production blasts according to a given fragmentation

The production cycle of an exploitation site generally develops in two phases: excavation and processing. The excavation technique depends on both the type of deposit and the geomechanical characteristics of the exploited material. The drill and blast technique involves carrying out a series of cyclical operations, consisting of evaluation of the geometry of the blast, choice of explosives and initiation systems, charging and initiation, scaling, loading, hauling, and dumping. Based on the different phases of the cycle, there are many factors that influence and are influenced by the size of the blasted material (Ouchterlony, 2003). Particularly, the grain size distribution and its maximum size "D_{max}" is important, as it represents the maximum acceptable size for the opening slot of the jaw crusher. Therefore, the desired fragmentation is a key parameter in the design of a blast (Fourney, 1993; Hustrulid, 1999; Katsabanis and Omidi, 2015); it is influenced by the drilling diameter, the blasting pattern, the type, and the amount of charge used (Onederra et al., 2004). To study the fragmentation induced by a blast, the geomechanical properties of the rock mass and its behavior under the explosives must be considered: as a first approximation, the impedance of the "rock-explosive" pair is a parameter that allows a priori estimation of the blast result in a given rock mass (Lu et al., 2016). The size (particle size distribution) of a blast mainly depends, for the same rock/explosive pair, on three factors:

- natural distribution of discontinuities (large blocks will never be obtained from a very jointed rock);

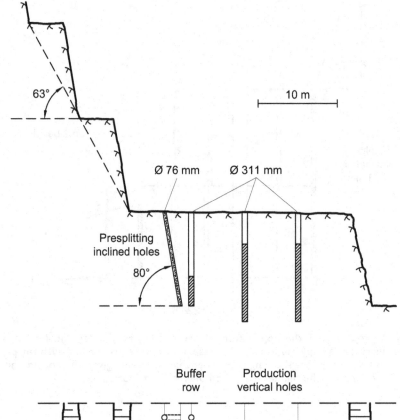

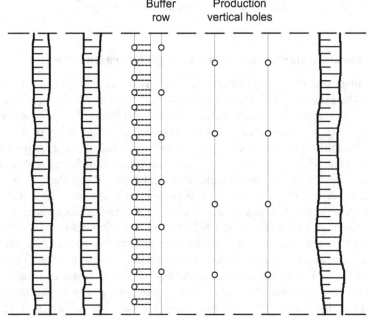

Figure 6.25 An example of the application of the presplitting with inclined holes for profiling the walls of a quarry at the end of the exploitation. Notice the charge reduction of the last production row (buffer row) improving the effectiveness of the pre-splitting.

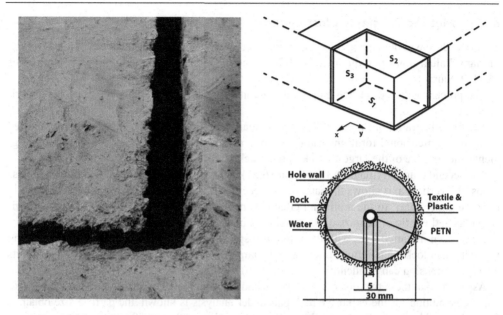

Figure 6.26 Left: A granite block (500 m³) separated from the solid and displacement by approx. 0.3 m by a detonating cord split blasting. The example represents a comparatively small blast; primary cut operations usually involve volumes in the range of 1,000–10,000 m³; right: above, scheme of the detonating cord splitting blast, with three splitting surfaces; below typical cross section of a charged hole.

- the blasting pattern (unless extremely weak charges are used, blocks larger than the pattern will never be obtained in the muck pile);
- the specific consumption (by increasing the charge, the grain size distribution obtained will be finer, provided the rock and blast geometry remains unchanged).

Of course, the particle size can be in some way checked through the two project variables (the pattern and the powder factor) only when massive or slightly fractured rocks must be excavated: in highly fractured rocks, the grain size is always ruled by the spacing of the discontinuities.

Furthermore, the practical rules provided are valid for systematic blasts: even though not specified, they implicitly consider the statement that the blast does not act on intact rock, but on a rock that has already been affected by previous blasts.

The rules are based on two assumptions:

- There is a minimum effective powder factor (P.F.$_{min}$) that is known in some way, representing the minimum value below which the fragmentation cannot occur, and immediately above which some fragmentation and dislocation can be obtained.
- The type of equation describing the particle size distribution of the material, whether expressed analytically or graphically, is invariable. Note the reference to the type of equation (linear, power, exponential, etc.), not to the equation: in other words: a grain size distribution can be defined unless a certain number of parameters is unknown.

6.6.1   Method for sizing blasting in limestone quarries

The procedure was developed by Mancini and Occella (1964), from the analysis of blasts results in many Italian limestone quarries for different applications, using medium-power explosives (ANFO, Slurry).

The process is based on two empirical diagrams, obtained from the analysis of the experimental cases.

The first (diagram I of Figure 6.27) is the standard particle size obtained from a blast, reported in the non-dimensional form: instead of the dimensions, the ratio between the size of the fragments and the size of the largest block in the muck pile is reported on the horizontal axis.

The second diagram (diagram II of Figure 6.27) is the correlation between two dimensionless ratios: the vertical axis shows the ratio of the specific consumption adopted, or to be adopted in the case under study, to the minimum effective specific consumption (P.F.$_{min}$) and, on the horizontal axis, the ratio of the size of the maximum block in the muck pile to the burden.

It has been assumed that, with the minimum specific consumption still capable of breaking the rock, the blocks in the muck pile cannot be larger than the blasting pattern (if it is square, its average dimension can be identified with the burden V).

Assuming that these two diagrams are valid, and since the value of the minimum effective specific consumption for the explosive/rock pair under analysis is known, the particle size obtained from a blast with a known burden and specific consumption, or the specific consumption necessary once the burden has been selected to obtain a given particle size or, once the specific consumption has been established, the burden necessary to obtain a certain particle size can be predicted.

To solve the first problem, the ratio of the adopted specific consumption to the minimum effective specific consumption, both known, is calculated and, from this ratio, through diagram II, the ratio of the size of the maximum block to the average burden; since the latter is known, the size of the maximum block is easily calculated from the ratio. The latter is then used as a scale in diagram I, to obtain the expected percentages of smaller or larger blocks of a given size.

To solve the second problem, the scale of the diagram I must be identified, that is, the value of the maximum size that corresponds to the wanted percentage of larger or smaller blocks of a given predetermined value.

Example: if the goal is to obtain 75% of blocks smaller than 25 cm, by outlining the 75% mark from the vertical axis to the curve in diagram I, the ratio D/D_{max} corresponding to that percentage can be read on the horizontal axis; by dividing the value of the considered size (25 cm) by the ratio D/D_{max} (0.3), D_{max} is found (in this example, 25/0.3 = 83 cm).

If the average burden is known (e.g., V = 1.5 m), the ratio D_{max}/V can be easily found: 83/150 = 0.55.

By indicating this value on the horizontal axis (0.55) in diagram II, the ratio P.F./P.F.$_{min}$ can be evaluated on the vertical axis, that is, by how much the specific consumption must exceed its minimum effective value (in the example, 1.16 times). The minimum effective specific consumption (e.g., 180 g/m³) is multiplied by this value and the necessary specific consumption is obtained (in the case, 180 × 1.6 = 288 g/m³).

To solve the third problem, having found D_{max} as in the previous case, the number of times the established value of the specific consumption exceeds its minimum effective value can be found, entering graph II with this ratio (on the vertical axis), and the corresponding ratio between D_{max} and the side of the burden is read: by dividing the maximum size by this ratio the burden (or the average value of the pattern) is found.

In limestone rocks, the value of the minimum effective specific consumption varies between 150 and 200 g/m³.

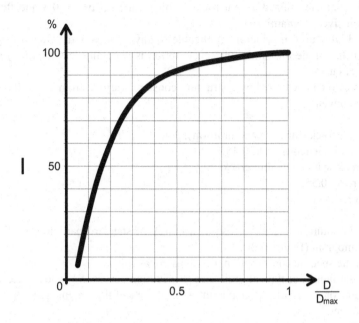

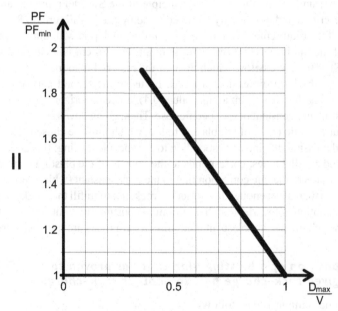

Figure 6.27 Nomograms for the calculation of specific consumption (redrawn from Mancini and Occella, 1964).

6.6.2   The SweDeFo model

The Swedish Detonics Research Foundation (Persson and Holmberg, 1983) has developed a model that considers the type of rock mass and the geometry of the blast. It assumes a family of "typical" particle size distributions for the muck pile and of a minimum specific consumption, characterizing different rocks, which varies between 0.3 and 0.5 kg/m³. The reference explosive is dynamite.

The method, although in principle applicable to any rock, was developed in particular for hard rocks (in limestone the minimum consumption is generally less than 0.3) and therefore suggests higher charges.

The rock is also characterized by a structure constant, varying from 0.4 to 0.6, according to the following conventions:

- Homogeneous rock (massive and uniform): 0.4.
- Relatively homogeneous rock: 0.45.
- Ordinary rock, with some thin cracks: 0.5.
- Fractured rock: 0.55.
- Very fractured rock: 0.6.

The method, in addition to making use of a family of standard particle size distributions, is based on a nomogram (Figure 6.28).

To find out the specific consumption, the steps are as follows:

Thanks to the family of typical grain size distribution curves, S_{50} is identified (average size of the fragments or, alternatively, maximum size of 50% of the fragmented material); this size is expressed in cm.

In order to better understand the working principle of the SweDeFo model, an example is in the following described and graphically depicted in nomogram of Figure 6.28.

For example, if the requirement is to have 80% of the muck pile with dimensions less than 70 cm, the curve of the family passing through the point 80%/70 cm is identified: this curve corresponds to an $S_{50} = 0.4$ (size below which 50% of the material is found = 40 cm).

By knowing S_{50}, the blast dimension proceeds by using the second part of nomogram (B of Figure 6.28). Entering the nomogram at this value (0.4), proceed horizontally up to the line corresponding to the structure constant of the specific case. Then, going upwards, the line corresponding to the spacing/burden ratio (E/V) of the blast is found, which must be chosen. Therefore, proceed horizontally until crossing the line corresponding to the specific drilling: this size, in blasts with a regular pattern and parallel holes, corresponds to the inverse of the blasting pattern area. Finally, continuing downwards to the line corresponding to the rock constant (which describes the amount of charge in a round that is just enough to extract the rock, not to fulfill the breakage requirements), the specific consumption is found. Note that minimum consumption values below 0.33 kg/m³ are not considered, which means that blasts in medium-strength rocks are not considered.

6.6.3 Transposition of a blasting plan that has proven to be valid for a given rock–charge pair to another rock–charge pair

The concept can be summarized as follows:

- Generally, a given blast is performed according to a scheme, optimized by a series of attempts, working on a certain rock/explosive pair.
- If the rock changes, or if the explosives change, or if both conditions occur, an attempt is made to adjust the pattern tested by use to the new situation, obviously with small changes (width of the pattern, diameter, within the compatible limits of the equipment available).

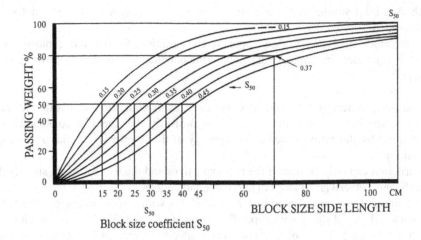

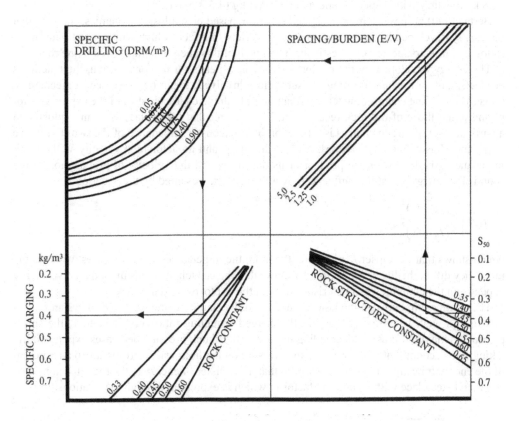

Figure 6.28 Nomograms for calculating the charges according to the Swedish method (Sandvik – Tamrock Corp., 1999). A: typical grain size curves; B: a nomogram for the calculation of the specific consumption.

- In short, a blast similar to the one already in use is redesigned, but with another specific consumption.
- It is, therefore, necessary to find a specific consumption, for the new explosive/rock pair, which is equivalent in terms of efficiency to the one that was used with the other explosive/rock pair.

The most basic criterion is to consider two charges as equivalent when they develop the same explosion energy and then, simply, to multiply the specific consumption of the scheme in use by the ratio of the specific energy of that explosive to that of the explosive that must replace it.

This criterion can only be used if the two explosives and the two rocks are slightly different from each other; otherwise, it leads to unreliable indications.

For example, it can be assumed that 1 kg of slurry with a specific energy of 3.5 MJ/kg is equivalent to 1.16 kg of slurry with a specific energy of 3 MJ/kg (3.5/3 = 1.16), but it is certainly incorrect to assume that 1 kg of dynamite with a specific energy of 4.5 MJ/kg is equivalent to 1.28 kg of slurry with a specific energy of 3.5 MJ/kg (4.5/3.5 = 1.28).

Berta (1990) proposes a more weighted criterion, based on explosion dynamics, showing that the energy developed in detonation produces a variety of effects which will be dealt with in the following. It is first necessary to estimate how much of this energy is transferred to the rock.

The energy developed by the explosive reaction depends on its composition, the reaction's products, and on the heat forming the substances involved; it can be expressed in mechanical units (MJ/kg). The energy transfer is a function of both the characteristics of the explosive supplying it and those of the rock receiving it. The transfer that takes place with an explosive in a particular set of circumstances is a function of the acoustic impedance of the two. Explosive impedance I_e is defined as the product of its density ρ_e and its detonation velocity VOD. Rock impedance I_r is defined as the product of its density ρ_r and the velocity of propagation of the sound. The energy transfer is influenced by a factor η_1, represented by the equation:

$$\eta_1 = 1 - \frac{\left(I_e - I_r\right)^2}{\left(I_e + I_r\right)^2}$$

which shows that the closer I_r and I_e are, the more the impedance factor increases, whereas the more they differ, the lower the factor. It follows that the capacity to transmit and receive energy depends on the combined characteristics of both the explosive and the rock.

The relationship between the hole diameter and the charge diameter (D_f/D_c) plays an important role as well: when the hole is ideally charged (coupling ratio very close to 1) the shock pressure is maximum; as the decoupling increases, the shock pressure decreases exponentially: taking into account that a reduction of the pressure on the hole wall corresponds to a reduction of the energy transmitted, it can be deduced that the transfer of energy to the rock in a non-ideal situation takes place with a coupling factor η_2 which is expressed through the relationship:

$$\eta_2 = 1 - \frac{1}{\left[e^{D_f/D_c} - (e-1)\right]}$$

It shows how η_2 tends to the maximum when the coupling tends to 1 and rapidly decreases when the decoupling increases.

Therefore, according to this criterion, two charges would be equivalent if the products of the mass of the charge for the specific energy and for the product of the two efficiencies are almost the same.

This equivalence, of course, is valid only when operating on the same rock or on slightly different rocks.

6.6.4   Demolition of masonry

The demolition of masonry can accidentally occur in excavation works. It is generally developed with blasts of small, inclined holes, arranged in several rows.

Figure 6.29 and Table 6.6 show an example and the recommended charges.

The demolition of blocks, which are preferably carried out without explosives, are of course not "designed". They can be performed with in-the-hole charges and stemming, in-the-hole charges without stemming, simply resting on the artifact to be demolished and covered with earth, as shown in Figure 6.30.

The system to be favored is the first; the other two are feasible only in insulated areas because they often involve fly rock and much greater atmospheric overpressures.

The explosive used is generally dynamite. Specific consumption is very low in the first and second techniques (100 g/m³ or less), due to a large number of free surfaces of the target; they are quite high, of the order of 500 g/m³, with the simply resting charge method.

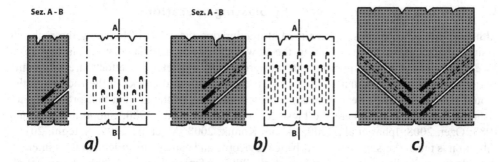

Figure 6.29 Demolition of a wall with rows of sideward holes: (a) with two rows of holes; (b) with three rows of holes; and (c) with three rows of holes on both sides

Source: Seguiti (1969)

Table 6.6 Elements for the demolition of walls using in-the-hole charges.

Wall thickness	Rows of holes	Diameter of the holes	Spacing between the holes of the same row	Spacing among the rows	Charge/ hole (1)
cm	n.	mm	cm	cm	g
30	2	22	23	23	45
45	2	22	30	30	90
60	2	32	46	46	120
90	2	32	60	60	240
120	3	32	76	76	460
180	3	36	91	91	900

Source: Seguiti (1969)

Note: (1) – The charges are valid for brick walls and must be increased when they are long or loaded; in addition, a 10% increase must be made for good masonry walls

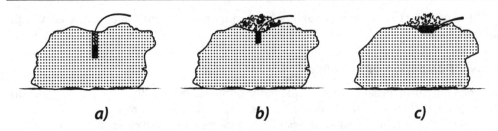

a)   *b)*   *c)*

Figure 6.30 Reducing the size of a block with a cartridge of explosive: (a) placed at the bottom of a stemmed hole; (b) introduced into a hole and covered with clay or sand; and (c) resting on the block and covered with clay or sand

Source: Seguiti (1969)

For these types of interventions, the explosive is applied anyhow quite rarely, as there are valid substitutes, such as chemical demolition agents, commonly applicable for the selective demolition of unit masonry (Natanzi and Laefer, 2014; Natanzi et al., 2020).

6.7   Ground vibration induced by blasting operations

Although the use of explosives is aimed at maximizing the effect of breaking the rock into desired shapes and sizes, controlling the unwanted effects associated with blasting to an acceptable limit is also important; the most remarkable deleterious effects are the launch of fragments at a great distance (fly rocks), the propagation of cracks in the rock mass that should remain in place after the blast, and the vibrations of the ground that propagate at a distance from the point where the blast occurs (Siskind et al., 1980; Shoop and Daemen, 1983; Ozer, 2008; Iphar et al., 2008, Ak and Konuk, 2008; Ak et al., 2009; Nateghi, 2011;). Vibration is the major source of disturbance to people and damage to structures if its intensity exceeds the safe vibration limits (Ozer et al., 2008; Afeni and Osasan, 2009; Kostić et al., 2013; Cardu et al., 2012, 2019).

While the abolition or reduction of fly rocks or crack propagation is achievable with the correct design and execution of the blast, the ground vibration can only be contained by limiting the charges, that is, by introducing an additional condition in the design.

The effects of ground vibration are sometimes referred to as seismic effects; although they differ considerably in oscillation amplitude, frequency, and duration compared to the seismic events, and being, therefore, arbitrary to include them in the intensity scales of natural earthquakes, they can be described in a formally similar way, assimilating the motion of each element of the ground (particle) to the composition of three damped oscillatory motions, conventionally indicated as vertical, radial, and transverse (see Figure 6.31).

For simplicity, for such motions, a sinusoidal trend is assumed: it can be in fact completely described with only two parameters chosen from the following four:

• the maximum value of particle velocity;
• the maximum value of particle acceleration;
• the maximum value of elongation;
• the frequency of oscillation.

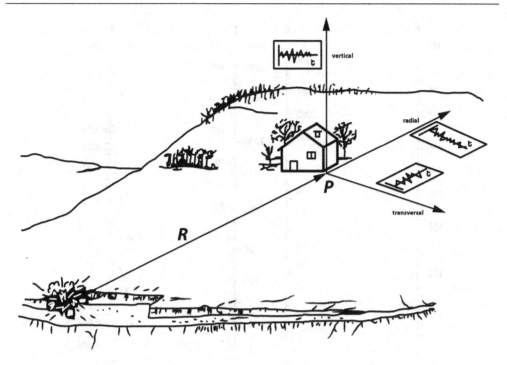

Figure 6.31 Vibrations induced in point P by a charge Q, exploding at a distance R, described by three-component vibratory motions, namely vertical (v), transverse (t), and radial (r).

For example, if the maximum value of particle acceleration and elongation is known, the maximum value of velocity and frequency can be obtained, etc.

The simple nomogram shown in Figure 6.32 (Langefors and Kihlström, 1967) allows obtaining, once two of the parameters are known, the remaining two.

Furthermore, in most cases, the parameters of the three-component motions (which are detected by the three sensors of the seismograph) are not used, but those of the resulting motion. In this way, the phenomenon can be described with only two numbers; the criterion commonly used to judge the harmfulness of vibrations is based on two values: the peak particle velocity (PPV), usually expressed in mm/s, and the frequency, expressed in Hz (s^{-1}). The harmfulness increases with the increase of the PPV and with the decrease of the frequency.

Figure 6.33 shows the tolerated ranges of vibrations induced by blasts for different types of structures, suggested by the German standard DIN 4150 (1984), which is usually referred to in Italy.

These limits have been established with a very wide margin of safety and do not refer to the possibility of structural injury, but rather to cosmetic damages, such as cracks in the plaster. For very sensitive and/or historical buildings, for example, ancient monuments, lower limits can still be imposed (Konon and Schuring, 1985; Jordan et al., 2009; Singh and Roy, 2010).

The ground vibrations are induced by the detonation of the charge through a rather complex mechanism. To elaborate more, four physical zones or environments associated with the

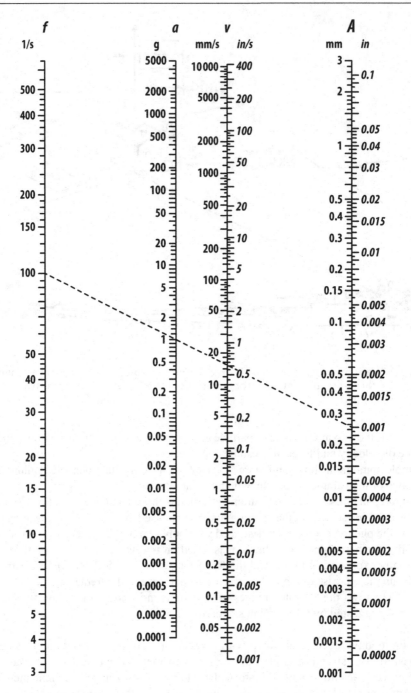

Figure 6.32 Nomogram (Langefors and Kihlström, 1967) for the graphic calculation of two of the vibration parameters when the others are known. The parameters considered are the frequency "f", in Hz, the maximum acceleration "a", expressed dimensionally as a multiple or sub-multiple of the gravity acceleration "g", the maximum particle velocity "v", expressed in in/s (left scale) or in mm/s (right scale), the maximum elongation "A", mm (left-hand scale) or in (right-hand scale). The nomogram is applicable to sinusoidal vibrations or to vibrations approximating them.

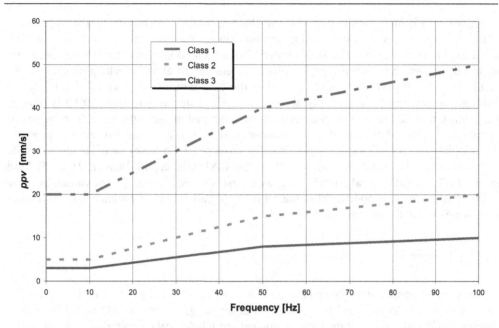

Figure 6.33 On the vertical axis: PPV, mm/s; on the horizontal axis: *f*, Hz. Trends 1, 2, and 3 show the limits provided by DIN 4150 for sensitive buildings (3), ordinary buildings (2), and industrial buildings (1), respectively. Reference is made to the PPV measured at the foundations.

generation, propagation, and recording of seismic blast vibrations can be defined: the generation zone, the seismic transmission zone, the acoustic transmission zone, and the recording site (Bollinger, 1980). It is within the generation zone that the rapid release of chemical energy, that is the explosion, takes place. Very high pressures and temperatures are generated within a very short time interval. This process causes the surrounding rocks to melt, flow, crush, and fracture. At some distance from the explosion site, these inelastic processes cease until the elastic effects begin; only a portion of the original chemical energy is converted to this elastic form. Naturally, the elastic disturbance that propagates away from the blast area is due to the seismic waves. The generation zone is defined, then, as the volume bounded by the elastic-inelastic interface. The seismic transmission zone is characterized by elastic effects, that is, it is where the solid medium returns to its original configuration after the passage of the seismic disturbance. In addition to inducing disturbance to the surrounding rocks, it happens very often that some of the explosion energy is transferred to the atmosphere (acoustic transmission zone, which can be the source of considerable public relations difficulties). As for the recording site, pertinent effects to be considered are ground-seismograph coupling and overburden effects.

The seismic speed depends on the type of rock and is within the order of a few km/s (of course, it has nothing to do with the particle velocity).

The intensity of the vibrations detected at a given distance from the blast depends on the distance itself, the amount of charge, and the nature of the rock, according to a general law of the type:

$$v = K \times R^{\alpha} \times Q^{\beta}$$

where v is the peak particle velocity, R is the distance, Q is the amount of charge, K, α, and β are experimental constants dependent on the types of blast and the type of rock (α

is negative). It is important to note that the charge referred to is not the total charge of the blast, but the amount of it detonating simultaneously: for example, if the blast uses 100 kg of explosive, yet distributed in ten holes loaded with 10 kg that detonate in succession, with a delay time between the explosions large enough to avoid overlapping effects, the result in terms of particle velocity is due to the contribution of 10 kg of charge, not to 100 kg. For this reason, commonly adopted formulas refer to Q in terms of CPD (charge per delay), that is, the sum of charges that are triggered at the same time. The ranges of commercial series of micro-delayed detonators (20–30 ms) are generally suitable to avoid overlapping effects. It was also empirically observed (Ambrasey and Hendron, 1968; Attewell et al., 1965; Duvall et al., 1963; Foster, 1981; Ghosh and Daemen, 1985; Siskind et al., 1980; Tripathy et al., 1995) that, as a first approximation, the maximum particle velocity is inversely proportional to the ratio $R/\sqrt{Q}$, called *scaled distance*. Therefore, the equation is given the practical form (site law):

$$v = K\left(\frac{R}{\sqrt{CPD}}\right)^{-n}$$

If v (maximum particle velocity or peak particle velocity -PPV) is given in mm/s, R, the distance, in m, and CPD, the maximum charge per delay, in kg, the two site constants take values, functions of the site, that is of the local geological and lithological characteristics as well as the type of blast used, widely variable (100–1,000) for the first (K) and a little less variable (from 1.5 to 1, but it is an exponent!) for the second (n).

Table 6.7 shows a collection of site constants measured in several open-cast operations in a dozen locations. It is easy to realize both, the great variability from case to case of the constants and the fact that there is no apparent correlation between them, with some definable characteristic of the medium, not to mention the lithological designation: it is not possible to deduce the values of the constants from the description of the rock alone and, in case of works in the vicinity of inhabited areas, it is practically inevitable to resort to blast tests and to the detection, at different distances, of vibrograms, to obtain the constants experimentally, unless they are already known from measurements made during similar works previously carried out in that area.

Even to set up a blast tests campaign, it is however necessary to get some idea of the values that the PPV will take at the measuring points if only to set up the full scale and the activation threshold of the tools. For this purpose, it can be assumed $K = 500$ and $n = -1.5$.

The values of K and n depend heavily on the type of work, as well as on the rock.

Blasts for tunnel excavation, with the same type of rock and CPD, give rise to more severe vibrations than those for open-pit excavations. Atmospheric overpressure (impulsive sound pressure) can also cause problems; it depends as well on the charge and the distance, according to a site law which is usually expressed as:

$$P_s = K_s \cdot \frac{Q^{1/3}}{R}$$

where P_s is the sound pressure, Q is the charge, and K_s is a constant to be experimentally found by means of tests, being even more variable than the constant relative to the ground vibration: it depends on the local topography, the presence, or absence, of natural or artificial shielding and on the way the blast-holes are arranged, loaded, stemmed, and detonated.

Table 6.7 Examples of site constants that were measured in several cases where drill and blast was adopted.

Location	Rock	Compressive strength	Tensile strength	Shear strength	Density	Young's modulus	Poisson's ratio	Site constants	
		MPa	MPa	MPa	kg/m3	Gpa		K	n
Lambidhar Mining Project Mussoorie	Gray marble	73.33	6.96	10.19	2,480	-	-	129.31	-1.39
Karankote limestone	Limestone	214.02	10.58	19.80	2,660	37.0	-	154.28	-1.38
Quarry, Tandur, CCI	Limestone	149.61	14.61	18.82	2,670	37.0	-	65.03	-0.68
Nayagaon limestone quarry, CCI	Sandstone	78.14	4.31	-	2,560	16.66	-	360.6	-1.89
Damodar OCP, Sudamdih, BCCL	Limestone	127.94	8.92	20.19	2,680	18.2	-	162.97	-1–15
Yerraguntala limestone quarry, CCI	Sandstone	139.10	9.15	-	2,500	-	-	181.56	-0–97
Bhimsagar Dam, Rajasthan	Granite	141.11	6.62	-	2,500	-	-	483.02	-1.56
North Koel Hydroelectric Project, Bihar	Sandstone	30.00	3.26	-	2,340	-	-	286.34	-1.25
Weat Mudidih, OCP, BCCL	Sandstone	18.92	2.47	-	2,200	-	-	292.45	-1.5
Gopinathpur, OCP, ECL	Sandstone	33.33	9.31	-	2,000	-	-	324.79	-1.23
Cargaly Colliery, quarry n. 3, ECL	Dolomite	96.08	-	-	2,860	40.7	0.35	174.58	-1.51
Salal hydroelectric project, J&K	Basalt	121.50	-	-	2,700	-	0.21	13.39	-0.76
Narmada Hydroelectric project, MP									

Source: Adhikari and Gupta (1989)

The vibration frequency of the ground (which is a parameter on which the permissible PPV values depend) also varies with the distance from the blast, decreasing as it grows. The following empirical law (Medvedev, 1963) can be used to describe this dependence:

$$F = \left(K_{\mathrm{f}} \cdot \log R\right)^{-1}$$

where F is the frequency and K_{f} an experimental site constant that generally varies in the range 0.01–0.03.

This empirical law, as can be easily understood, cannot be valid at small distances, because it would provide unreliable results: at 1 m distance, it would give infinite frequency, and at less than 1 m, it would give negative frequency. It provides reasonable interpolation at distances of tens or hundreds of meters; however, in the case of small blasts, only the PPV, in practice, is of concern (the tolerable limit value must be lowered as the frequency decreases; at frequencies above 100 Hz, the standards prescribe a maximum limit of PPV independent on the frequency).

Also, the K_{f} constant of the formula must be determined with measurements in specific sites.

In conclusion, the steps to be taken to solve the problem of vibrations, limited to the technical part, were summarized; the non-technical aspects, namely the creation of an atmosphere of mutual clarity, trust, and cooperation between the Company, third parties, and local authorities are perhaps more important, but nothing can be suggested in this context.

In case of explosions close to buildings that could be damaged by vibrations, it is necessary to plan the first inspections, locate them on the map, and accurately describe their nature and state, mainly for three reasons:

- Buildings can be more or less "sensitive". It is different to organize a blast at 150 m from a sturdy industrial building in reinforced concrete, or from a brick house. Apart from these extreme cases, the classes shown in Figure 6.33 are rather vaguely defined and the attribution of a building to one class or another cannot be made by "hearsay".
- Anyone who owns a car knows how irresistible it is to be tempted, when hit by another vehicle, to take responsibility for all visible dents on it, even if the impact did not cause any damage. Against this human weakness, the driver has no defense, because he cannot have an updated report on the status of each of the millions of vehicles that chance brings him to meet. The company that works with explosives is in a potentially better position: there aren't a million buildings nearby, and they don't move.
- The Insurance Company generally requires the execution of inspections and the description of the nature and state of the surrounding structures, for obvious reasons.

It is then necessary to establish the values to be assigned to the constants of the site: these can be known from other works in the area, or can be determined with cautious blast tests: the vibrations are detected at different distances from the blast and the experimental values are reported on log/log paper, with the values of V in the vertical axis and the values of scaled distance in horizontal axis, as shown in Figure 6.34.

It is easy to notice that in log/log coordinates, the site law is represented by a straight line.

This line is drawn by interpolation; from its intercept with the vertical axis and from its slope, the values of K and n are obtained, respectively. From the same surveys, the value of the predominant frequency is also found.

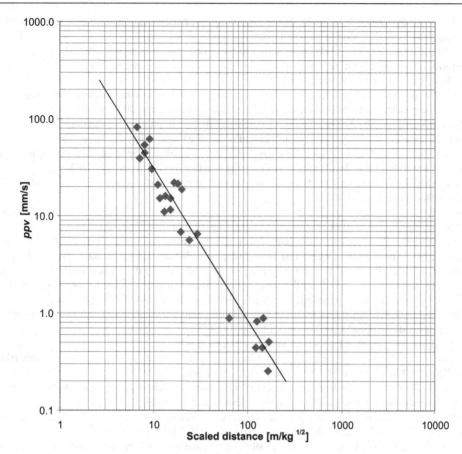

Figure 6.34 Diagram of PPV values as a function of the scaled distance: example of interpolation and definition of the site law; in the specific case, $ppv = 1076.6 \cdot \left(\frac{D}{\sqrt{cpd}} \right)$, with a mean-square deviation close to unity (0.95) and a standard error of 0.18, confirming the poor dispersion of the data (overall 25 points) and therefore the reliability of the interpolation.

Once the site law constants are known, the PPV values that must not be exceeded are determined, for example, based on the limits shown in Figure 6.33 and of what has been acquired on the buildings to be protected. Since the distances of these buildings from the predicted blast (R) are known from the map, as well as the site constants (K and n) and the speed limit value (v), the site law is solved for CPD.

The lowest of the CPD values detected becomes a condition to be respected in the design of the blast.

References

Adhikari, G.R., Babu, A.R., Balachander, R. and Gupta, R.N., 1999. On the application of rock mass quality for blasting in large underground chambers. Tunnelling and Underground Space Technology, 14(3), pp. 367–375.

Adhikari, G.R. and Gupta, R.N., 1989. Some aspects of cast blasting techniques for dam construction. International Journal of Mining and Geological Engineering, 7(4), pp. 301–313.

Afeni, T.B. and Osasan, S.K., 2009. Assessment of noise and ground vibration induced during blasting operations in an open pit mine – a case study on Ewekoro limestone quarry, Nigeria. Mining Science and Technology (China), 19(4), pp. 420–424.

Ak, H., Iphar, M., Yavuz, M. and Konuk, A., 2009. Evaluation of ground vibration effect of blasting operations in a magnesite mine. Soil Dynamics and Earthquake Engineering, 29(4), pp. 669–676.

Ak, H. and Konuk, A., 2008. The effect of discontinuity frequency on ground vibrations produced from bench blasting: A case study. Soil Dynamics and Earthquake Engineering, 28(9), pp. 686–694.

Ambrasey, N.R. and Hendron, A.J., 1968. Dynamic behavior of rock masses. K.G. Stagg and O.C. Zienkiewicz (Eds.), Rock Mechanics in Engineering Practice. Wiley, London, pp. 203–227.

ATLAS COPCO, 2007. Mining Methods in Underground Mining – Atlas Copco Rock Drills AB. Atlas Copco, Stockholm, pp. 33–45.

Attewell, P.B., Farmer, I.W. and Haslam, D., 1965. Prediction of ground vibration from major quarry blasts. International Journal of Mining and Mineral Engineering, Vol. 5, pp. 621–626.

Barnes, J.W., 1947. Development of the demolitions tape. The Military Engineer, 39(260), pp. 248–250.

Berta, G., 1990. Explosives: An Engineering Tool. Italesplosivi, Milano. Ed. La Moderna, Novara, Italy.

Billaux, D., Chiles, J.P., Hestir, K. and Long, J., 1989. Three-dimensional statistical modelling of a fractured rock mass – An example from the fanay-augeres mine. International Journal of Rock Mechanics and Mining Sciences & Geomechanics Abstracts, 26(3–4), July, pp. 281–299. Pergamon.

Boadu, F.K., 1997. Fractured rock mass characterization parameters and seismic properties: Analytical studies. Journal of Applied Geophysics, 37(1), pp. 1–19.

Boky, B., 1967. Mining. MIR Publishers, Moscow, RU, pp. 54–56.

Bollinger, G.A., 1980. Blast vibration analysis. Chapter II: The Generation of Seismic Waves from Blasting. Southern Illinois University Press, Carbondale, IL, USA.

Bowa, V.M., 2015. Optimization of blasting design parameters on open pit bench a case study of Nchanga open pits. International Journal of Engineering, Science and Technology, 4(9).

Büsch, J., 1977. Sprengen mit Hintereinander Gestaffelten, von einer Reihe aus mit Unterschiedlichen Neigungen Hergestellen Großbohrlöchern. Nobel Hefte, 43(3), pp. 96–100.

Cardu, M., 1990. L'abbattimento con cariche sferiche e la sua applicazione nel metodo di coltivazione VCR. Bollettino della Associazione Mineraria Subalpina, 27(4), pp. 661–693.

Cardu, M. and Castelli, E., 1996. Ottimizzazione dei metodi di coltivazione in cave di porfido del Trentino. Atti IV Congresso Italo-Brasiliano di Ingegneria Mineraria, Canela-RS, Brasile, pp. 145–151.

Cardu, M., Castelli, E., Fornaro, M. and Mancini, R., 1995. Blasting in densely jointed rock to obtain slabs. 21st International Conference on Explosives and Blasting Technique, International Society of Explosives Engineers, Nashville, TN, USA, pp. 132–142.

Cardu, M., Coragliotto, D. and Oreste, P., 2019. Analysis of predictor equations for determining the blast-induced vibration in rock blasting. International Journal of Mining Science and Technology, 29(6), pp. 905–915.

Cardu, M., Dompieri, M. and Seccatore, J., 2012. Complexity analysis of blast-induced vibrations in underground mining: A case study. International Journal of Mining Science and Technology, 22(1), pp. 125–131.

Chernigovskii, A.A., 1986. Applications of Directional Blasting in Mining and Civil Engineering, Balkema, Rotterdam.

Chung, S.H. and Katsabanis, P.D., 2000. Fragmentation prediction using improved engineering formulae. Fragblast, 4(3–4), 198–207.

Clark, G.B., 1987. Principles of Rock Fragmentation. J. Wiley & Sons, New York, USA.

Crackel Jr., M., Heisel, M. and Ramos, G.G., 1981. Stope blasting design and experience at the carr fork mine. Proceedings of the International Conference on Caving and Sublevel Stoping, Denver, USA, pp. 529–533.

Del Greco, O., Fornaro, M., Mancini, R. and Patrucco, M., 1983. Profilatura dei fronti di cava nell'abbattimento con esplosivi: analisi di alcuni esempi notevoli. Bollettino dell'Associazione Mineraria Subalpina, 20(1–2), pp. 136–159.

Duvall, W.I., Johansson, C.F., Meyer, V.C.A. and Devine, J.F., 1963. Vibrations from blasting instantaneous and millisecond delayed quarry blasts. United States Bureau of Mines, Report of Investigations, No. 6151.

Foster, A.G., 1981. Structural response and human response to blasting vibration effects-is there any connection? Proceedings of the Conference on Explosives and Blasting Technique, Society of Explosive Engineers, Phoenix, AZ, USA, pp. 10–26.

Fourney, W.L., 1993. Mechanisms of rock fragmentation by blasting. Comprehensive Rock Engineering Principles, Practice and Projects, vol. 4. Pergamon Press, Oxford, UK, pp. 39–69.

German Standards Organization (GSO), 1984. Vibrations in building construction, DIN 4150, Berlin.

Ghosh, A. and Daemen, J.K., 1985. Statistics, a key to better blast vibration predictions. E. Ashworth (Ed.), Proceedings of the 26th U.S. Symposium on Rock Mechanics, Rapid City, SD, USA, pp. 1141–1149.

Gois, J.C., 2019. PECCS-pan European competence certificate for shot-firers: Blasting theory and design. EFEE, Shotfiring Committee Workshop, Coimbra, Portugal.

Hino, K., 1956. Fragmentation of rock through blasting and shock wave theory of blasting. The 1st US Symposium on Rock Mechanics (USRMS), ARMA-56-0191, Golden, CO, USA, April 23–25.

Hudaverdi, T., Kuzu, C. and Fisne, A., 2012. Investigation of the blast fragmentation using the mean fragment size and fragmentation index. International Journal of Rock Mechanics and Mining Sciences, 56, pp. 136–145.

Hustrulid, W.A., 1999. Blasting Principles for Open Pit Mining, Vol. 1: General Design Concepts and, Vol. 2: Theoretical Foundations. A.A. Balkema, Rotterdam, the Netherlands.

Iphar, M., Yavuz, M. and Ak, H., 2008. Prediction of ground vibrations resulting from the blasting operations in an open-pit mine by adaptive neuro-fuzzy inference system. Environmental Geology, 56(1), pp. 97–107.

Jordan, J.W., Sutcliffe, D.J. and Mullard, J.A., 2009. Blast vibration effects on historical buildings. Australian Journal of Structural Engineering, 10(1), pp. 75–84.

Kamarudin, N., Ghani, N., Mustapha, M., Ismail, A. and Daud, N., 2012. An overview of crater analyses, tests and various methods of crater detection algorithm. Frontiers in Environmental Engineering, 1(1), pp. 1–7.

Katsabanis, P.D. and Liu, L., 1998. A numerical study of some aspects of the spherical charge cratering theory. Fragblast, 2(2), pp. 219–233.

Katsabanis, P.D. and Omidi, O., 2015. The effect of delay time on fragmentation distribution through small and medium scale testing and analysis. Fragblast 11: Proceedings of the 11th International Symposium on Rock Fragmentation by Blasting, AusIMM, Carlton, Australia, pp. 715–720.

Konon, W. and Schuring, J.R., 1985. Vibration criteria for historic buildings. Journal of Construction Engineering and Management, 111(3), pp. 208–215.

Kostić, S., Perc, M., Vasović, N. and Trajković, S., 2013. Predictions of experimentally observed stochastic ground vibrations induced by blasting. PLoS One, 8(12), p. 82056.

Lang, L.C., 1982. Vertical crater retreat: An important new mining method. Underground Mining Methods Handbook. AIME, New York, pp. 456–463.

Lang, L.C., Roach, R.J. and Osoko, M.N., 1977. Vertical Crater retreat an important new mining method. Canadian Mining Journal, 98(9), p. 69.

Langefors, U. and Kihlström, B., 1967. The Modern Technique of Rock Blasting, 2nd Ed. Almqvist & Wiksell, Stockolm.

Livingston, C.W., 1956. Fundamental concepts of rock failure. Quarterly of the Colorado School of Mines, 51, 3, Denver, USA.

Lu, W., Leng, Z., Chen, M., Yan, P. and Hu, Y., 2016. A modified model to calculate the size of the crushed zone around a blast-hole. Journal of the Southern African Institute of Mining and Metallurgy, 116, pp. 413–422.

Lundborg, N., 1967. The strength-size relation of granite. International Journal of Rock Mechanics and Mining Sciences & Geomechanics Abstracts, 4(3), pp. 269–272. Pergamon.

Mancini, R. and Cardu, M., 1995. A static model for rock splitting design with explosive. Proceeding of the XXI Annual Conference on Explosives and Blasting Technique, Nashville, TN, USA, February 5–9, pp. 124–131.

Mancini, R., Cardu, M. and Ferrero, A.M., 1992. The detonating cord-cutting method in dimension stone quarries: Theoretical approaches, compared with field data. Proceeding of the III Geo Engineering Congress, Turin, Italy, December 1–2.

Mancini, R., Cardu, M. and Fornaro, M., 1994. Granite blocks extraction by detonating cord in different Italian quarry basins: An overview of design and operation practices. Proceeding of the XX Annual Conference on Explosives and Blasting Technique, Austin, January 30–February 3, pp. 305–316.

Mancini, R., Cardu, M. and Fornaro, M., 1995. An analysis of the influence of rock quality on the results of controlled blasting practices. Proceedings of the II Int. Conf. on Engineering Blasting Technique, Kunming (China), pp. 264–267.

Mancini, R., Fornaro, M. and Cardu, M., 1993. No-fragmentation blasting: The detonating cord quarrying method for dimension stones. Proceedings of the International Conference Fragblast-4, Rock fragmentation by blasting, Vienna, Austria, pp. 431–436.

Mancini, R. and Occella, E., 1964. Inquadramento statistico di lavori di abbattimento con esplosivi sulla base di un criterio di similitudine. Memorie e Note dell'Istituto di Arte Mineraria del Politecnico di Torino, Tip, Artigianelli, Torino.

Medvedev, S.V., 1963. The problem of investigations of the seismic effect of explosions at the institute of physics of the earth, U.R.S.S. Academy of Sciences. Problems of Engineering Seismology, Consultant Bureau, New York.

Merriam, R., Rieke III, H.H. and Kim, Y.C., 1970. Tensile strength related to mineralogy and texture of some granitic rocks. Engineering Geology, 4(2), 155–160.

Natanzi, A.S. and Laefer, D.F., 2014. Using chemicals as demolition agents near historic structures. 9th International Conference on Structural Analysis of Historical Constructions, Mexico City, Mexico, October 14–17.

Natanzi, A.S., Laefer, D.F. and Zolanvari, S.I., 2020. Selective demolition of masonry unit walls with a soundless chemical demolition agent. Construction and Building Materials, 248, p. 118635.

Nateghi, R., 2011. Prediction of ground vibration level induced by blasting at different rock units. International Journal of Rock Mechanics and Mining Sciences, 48(6), pp. 899–908.

Nedriga, V.P., Pokrovsky, G.I. and Lushnov, N.P., 1983. Directional blasting in rockfill dam construction. International Water Power and Dam Construction, 35(6).

Ninahua, Y. and Yongji, H., 1994. Problems of slope stability caused from blasting for dam construction. Water Resources and Hydropower Engineering, 22(3), pp. 389–398.

Olson, J.J., Dick, R.A., Condon, J.L., Hendrickson, A.D. and Fogelson, D.E., 1970. Mine Roof Vibrations from Underground Blasts. US Department of the Interior, Bureau of Mines, Shullsburg, WI, USA, RI 7330, 55 pp.

Olson, J.J. and Fletcher, L.R., 1971. Airblast-Overpressure Levels from Confined Underground Production Blasts (No. 7561–7580). US Bureau of Mines.

Onederra, I., Esen, S. and Jankovic, A., 2004. Estimation of fines generated by blasting – applications for the mining and quarrying industries. Mining Technology, 113(4), pp. 237–247.

Ouchterlony, F., 2003. Influence of blasting on the size distribution and properties of muckpile fragments, a state-of-the-art review. MinFo Project P2000-10: Energy Optimisation in Comminution, 114 pp.

Ozer, U., 2008. Environmental impacts of ground vibration induced by blasting at different rock units on the Kadikoy – Kartal metro tunnel. Engineering Geology, 100(1–2), pp. 82–90.

Ozer, U., Kahriman, A., Aksoy, M., Adiguzel, D. and Karadogan, A., 2008. The analysis of ground vibrations induced by bench blasting at Akyol quarry and practical blasting charts. Environmental Geology, 54(4), pp. 737–743.

Persson, P.A. and Holmberg, R., 1983. Rock dynamics. Proceeding of the 5th ISRM Congress, Melbourne, Australia, April, n. ISRM-5CONGRESS-1983-242.

Powell, H.H., 1948. Railroad turnout formulae. The Military Engineer, 40(271), pp. 212–215.

Rotzetter, G., 1977. Beseitigung der Bergsturzgefahr am Kirchberg in Meiringen (Schweiz) durch Sprengarbeit. Nobel Hefte, 43(2), pp. 101–106.

Sandvik – Tamrock Corp., 1999. Rock Excavation Handbook for Civil Engineering. Matti Heiniö Ed., Tampere, Finland, 364 pp.

Segui, J.B. and Higgins, M. (2002). Blast design using measurement while drilling parameters. Fragblast, 6(3–4), 287–299.

Seguiti, T., 1969. Le mine nei lavori minerari e civili. Editions of the Magazine "L'Industria Mineraria". P. Di Pietra Pub., Roma, Italy, 711 pp.

Sha, S., Rong, G., Tan, J., He, R. and Li, B., 2020. Tensile strength and brittleness of sandstone and granite after high-temperature treatment: A review. Arabian Journal of Geosciences, 13(14), 1–13.

Sheng, L.X. and Gu, Y, 1993. A successful directional blasting operation in the 10 kton class, to level the Zyhai Paotai Mountain, China. GEAM, XXX, 81(4), pp. 209–213.

Shoop, S.A. and Daemen, J.J., 1983. Site-specific predictions of ground vibrations induced by blasting. Preprints, Society of Mining Engineers AIME, Atlanta, GA, USA, Report n. 83(CONF-830317-), 83(95).

Singh, P.K. and Roy, M.P., 2010. Damage to surface structures due to blast vibration. International Journal of Rock Mechanics and Mining Sciences, 47(6), pp. 949–961.

Siskind, D.E., Stagg, M.S., Koop, J.W. and Dowding, C.H., 1980. Structure response and damage produced by ground vibration from surface mine blasting. United States Bureau of Mines, Report of Investigations, No. 8507.

Spannagel, C., 1969. Sprengtechnische Entwicklungen in der Südwestdeutschen Gevinnung von Gyps und Anidrit. Nobel Hefte, 31(1), pp. 19–25.

Starfield, A.M., 1966. Strain wave theory in rock blasting. The 8th U.S. Symposium on Rock Mechanics (USRMS), Minneapolis, MN, September 15–17. Paper Number: ARMA-66-0538.

Tripathy, G.R., Shirke, R.R., Marwadi, S.C. and Gupta, I.D., 1995. Attenuation characteristics of seismic waves generated due to blasting for rock excavation. Proceeding of the International Seminar on Rock Excavation Engineering, Present and Future Trends, A-II.1–II.12, Panjim, Goa, IN.

Wang, M., Qiu, Y. and Yue, S., 2018. Similitude laws and modeling experiments of explosion cratering in multi-layered geotechnical media. International Journal of Impact Engineering, 117, pp. 32–47.

Westwater, R., 1957. Heading blasting. Mine & Quarry Engineering, (7), pp. 292–298.

Wright, F.D., Burgh, E.E. and Brown, B.C., 1953. Blasting Research at the Bureau of Mines Oil-Shale Mine, vol. 4956. US Department of the Interior, Bureau of Mines, Rifle, CO, USA.

Xu, F., 1991. A throwing blasting CAD System. Proceeding of the International Conference on Engineering Blasting Technique (ICEBT), Peking University Press, Beijing, China, pp. 139–144.

Yang, X., Liu, C., Ji, Y., Zhang, X. and Wang, S., 2019. Research on roof cutting and pressure releasing technology of directional fracture blasting in dynamic pressure roadway. Geotechnical and Geological Engineering, 37(3), pp. 1555–1567.

Ye, T.Q., 2008. Field experiment for blasting crater. Journal of China University of Mining and Technology, 18(2), pp. 224–228.

Yue, Z.W., Yang, L.Y. and Wang, Y.B., 2013. Experimental study of crack propagation in polymethyl methacrylate material with double holes under the directional controlled blasting. Fatigue & Fracture of Engineering Materials & Structures, 36(8), pp. 827–833.

Chapter 7

Underground blasting

7.1 Production blasts in underground construction sites

The term *production blasts* refers to blasts that do not have as a goal the advancement in a blind-bottom excavation (tunnels, shafts, and raises) but the creation of large cavities aimed at exploiting the rock (mining purposes) or the cavity itself (civil works), after the necessary developments.

There is, of course, a huge variety of geometric configurations and sequences of intermediate steps to reach those goals; the typologies of production blasts include, in addition to the blasts for bottom-blind excavations (which will be examined in the following), other steps, that can be summarized as

- blasts of holes parallel to a free surface, in conditions of a constrained or free toe;
- blasts of inclined holes, contained in one or more planes, or cones, parallel to a free surface, with a bounded or free toe, drilled according to complete or partial rings;
- blasts of inclined holes arranged into fans, with an increasing inclination of the holes, in each fan, with respect to the free surface;
- blasts of spherical charges arranged into planes parallel to a free surface (crater blasting).

Apart from ring blasts (with full or partial rings) and the availability of free toes in some cases, the blasts are similar to those used in open-cast construction sites. The Powder Factor is systematically higher (1.5–2 times) than what is commonly adopted in an open-cast excavation with the same explosive-rock pair, also because the incidence of holes acting in constrained conditions (both laterally and at the toe) is often greater, as in a narrow trench.

Schematically, the work necessary to remove a large volume of a given geometry from the underground, without causing the immediate collapse of the rock above (in some mining operations, the landslide is expected as part of the production cycle, which cannot prescind from the real excavation work), can be split into the following parts:

1) performing the blind-bottom excavations (tunnels, raises, and inclined ways) to allow access to the volume to be blasted and/or to remove the blasted rock. This phase (commonly called "development") involves blasts with only one free surface and, therefore, very high unit costs and very low hourly productions. While preparing the work plan, efforts are made to minimize the impact of these auxiliary but necessary works on the total volume to be removed. The ratio of the length of blind-bottom excavations (expressed in m) to the volume of material that is intended to be exploited (expressed in m³) is called the development ratio

DOI: 10.1201/9781003241973-7

(expressed in m⁻²). It is an indicator used often to assess the success achieved in planning the work and to evaluate alternative solutions; of course, it is only an indicator, and rather vague, of the goodness of the solution, and must be weighed case by case. Usually, it varies between 0.1 and 0.01 (in some mining work, it drops below 0.01). Obviously, as a general rule, the lower this ratio is, the lower the cost of organizing the underground exploitation;

2) removing, thanks to the development works, suitably oriented volumes of rock (vertical, horizontal, or inclined slices) having dimensions suitable to provide the necessary drilling surfaces, or additional free surfaces, for the next phase (slotting and slashing);

3) implementation of the systematic production blasting for the removal of the residual, and prevailing, volume.

Figures 7.1–7.5 do not represent a real case, but only some possibilities; they do not suggest optimal solutions for a real case, and they do not exhaust the possibilities of solutions: they only offer the possibility of exemplifying numerous types of approaches for production blast, necessary before the rules for calculating charges are addressed.

The hypothetical objective is to create a cavity with a shape and size similar to those shown in the figures (it could be a volume of useful minerals for the exploitation or construction of an underground silo or reservoir, or others), composed of a cube with an edge length of 20 m below which a pyramid with a base square of 15 m of height is located.

Stability problems are not dealt with despite the necessity to be solved and serve only to examine the geometric aspects. Some of them present serious difficulties or practical contraindications, which are not discussed, but their analysis can be a useful topic for reflection.

The first solution (Figure 7.1) reduces the development operations to the completion of two tunnels and a connection raise. It does not require slotting operations, but simply drilling, loading (from a platform dropped in the raise), and then ring holes blasting in horizontal planes, increasing in length according to the evolution of the excavation, each one producing the

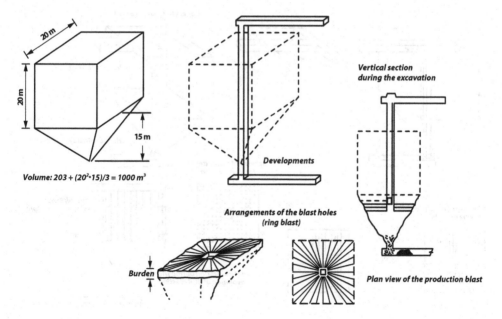

Figure 7.1 Underground production blasts: solution 1.

crushing and the collapse of a horizontal slice of rock; the blasted rock is then conveyed to the lower tunnel, from which it is extracted (c.d. Alimak method).

In the second solution (Figure 7.2), the development consists of a bottom tunnel, a top tunnel, and a connection raise; from the latter, the trunk-conical volume is removed operating as in the first solution. The top tunnel is slashed over the whole width of the cavity to be excavated, with horizontal fan blasts and, subsequently, with blasts of parallel horizontal holes. The excavated rock is discharged through the raise and taken from the base of the pyramidal cavity, which serves as a hopper. Alternatively, a T-bar can be drawn from the end of the top tunnel where the slashing is to be carried out, only with horizontal parallel blast holes. Then, at the level of the upper tunnel, a chamber with a plan corresponding to that of the cavity to be created is obtained. From this chamber, a pattern of vertical holes is drilled that reach the crown of the cavity. These holes are loaded from above with spherical charges, and placed at a suitable height with respect to the crown of the bottom slot so that a blast of spherical charges for crater blasting is organized, thanks to which a horizontal slice of rock is crushed and collapses in the hopper below. The operation is repeated with further blasts, removing other horizontal slices, until no horizontal pillar of enough thickness between the vacuum and the chamber remains to ensure the stability of the floor of the upper chamber. This pillar is also blasted according to slices but, of course, with a single blast of spherical charges organized by means of in-the-hole delayed decks between the charges.

In the third solution (Figure 7.3), the development consists of a bottom tunnel, a top tunnel, and one between the trunk-conical and the prismatic section of the cavity to realize a raise in axis with the future pyramidal hopper, connecting the base tunnel with the intermediate one, a second raise at the bottom wall of the future cavity, connecting the ends of the middle and the top tunnels, and two T-bars at the end of those tunnels.

The intermediate tunnel is slashed to its perimeter with blasts of horizontal holes parallel to the crossbar, creating a chamber that covers the horizontal section of the cavity to be created (bottom slot), and, operating from this chamber, the lower pyramidal volume is excavated with

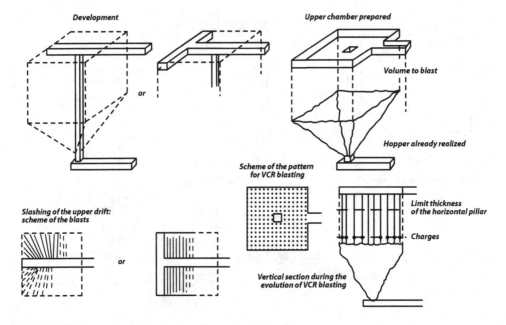

Figure 7.2 Underground production blasts: solution 2.

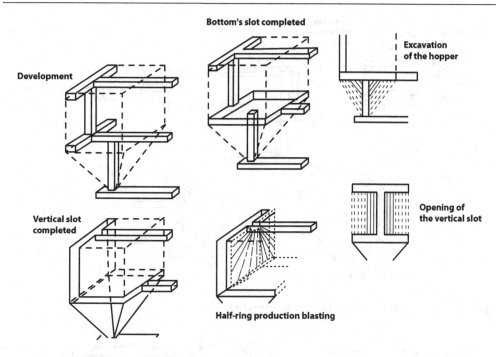

Figure 7.3 Underground production blasts: solution 3.

blasts of vertical holes which will be gradually inclined, until reaching the parallelism to the walls of the future hopper, using the central raise as the preliminary free surface.

Then, the vertical slot is realized, with blasts of parallel vertical holes, using one of the raises as a free wall that connects the intermediate tunnel to the top one. In practice, a trench is created along the sole of the bottom crossbar, starting from the upper level, with a depth corresponding to the difference between this sole and the roof of the bottom slot. The blasted rock descends to the base hopper, from where it is extracted.

At this point, the volume to be removed is isolated at the base and at the bottom and is blasted with semi-ring blasts in planes orthogonal to the top tunnel (and therefore parallel to the vertical slot), covering the whole excavation cross section, and drilled from the top tunnel itself. The blasted rock falls into the lower hopper, from where it is extracted (c.d. long hole sublevel or classic sublevel stoping).

In the fourth solution (Figure 7.4), the development is similar to that of the previous case, but the top tunnel is set to lower levels than that of the roof of the chamber to be excavated. The portion of the vertical slot necessary to reach the desired height of the room is obtained by slashing the crossbar for the necessary height, with blasts of inclined holes or fan holes (or a combination of the two) and the lower part as in the previous case, with vertical holes.

The removal of the volume, isolated from the two slots, is then performed with full ring blasts (this is a variant of the previous case; many others, not described, are possible).

In the fifth solution (Figure 7.5), the development and opening of the slots are similar to the third case, but the top tunnel is also slashed up to the excavation perimeter, possibly in more tranches. The removal of the volume is then performed with vertical blast holes parallel to the vertical slot. It is a kind of bench blasting, similar to those used for trenches, but without

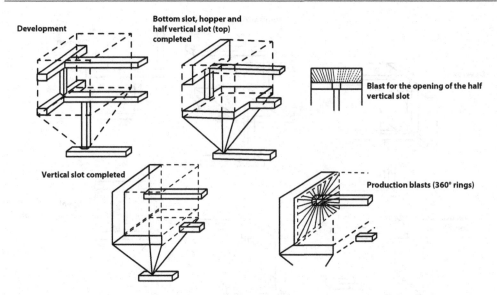

Figure 7.4 Underground production blasts: solution 4.

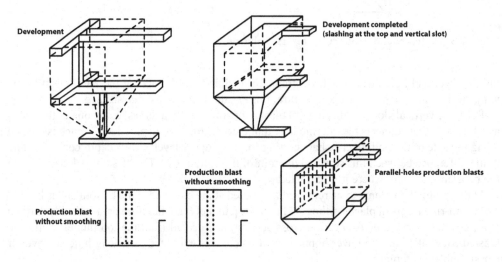

Figure 7.5 Underground production blasts: solution 5.

bounding the toe. To obtain a good regularity and integrity of the sidewalls, smoothing is also used in this case (thickening of the perimeter holes and reduction of their charge). The method is called big hole sublevel or parallel hole sublevel.

There are of course many other variants, or combinations of those described, but the examples provided are sufficient to clarify the possible types of blasts.

For sizing underground production blasts, both with parallel and inclined holes, the rule is to proceed in the same way as for the excavation of trenches, with increasing specific consumption. Often, the holes are very long and each requires a very large charge, up to hundreds of kg. A delay limit charge can be imposed, to avoid damage to the underground structure caused by vibrations (Zhang and Naarttijärvi, 2005; Soltani-Mohammadi et al., 2012; Ak et al., 2009), and in this case, the holes are often loaded according to decks, separated by layers of inert material of length sufficient to prevent flashover between the decks, and are triggered with detonators according to a predetermined delay scale. Even complete or partial ring blasts can be sized with the criterion of meeting predefined specific consumption values, with monomial or binomial formulas similar to those commonly adopted for parallel holes' blasts despite the different geometry (the purpose is to calculate volumes of wedges, rather than parallelepipeds). However, there is, of course, the problem of preventing the explosive from being in excess in some areas and missing in others: in fact, the distance between the holes is minimum at the mouth, maximum at the end, and if all the holes in the ring have been charged, for example by ¾ of their length, there would be a volume in the rock (that from the mouth of the holes to ¼ of their length) where there is no explosive, a second area (that, for example, from ¼ of the length to half-length) where there is an excess of explosive with respect to the volume to be blasted and a third, from half-length to the bottom of the hole, where both the charge and the volume to be blasted are fairly balanced with each other (Onederra and Chitombo, 2007; Crackel et al., 1981).

A similar problem also occurs in fan blasts, but these rarely cover surfaces comparable to those of ring blasts, and the problem can be solved by charging alternatively a hole over the whole length, except for the stemming, and the next one only half-length.

In the case of ring blasts, the problem is more serious: as an example, the case, not exceptional, of a ring drilled from a tunnel having a 4 m × 4 m cross section, with 10-m long holes, is considered. Such a ring covers an area approximately circular with a diameter of 24 m and a circumference of 75 m, while the perimeter of the tunnel from which the holes are drilled is only 16 m.

By giving the bottom of the holes a toe spacing of 1.5 m, suitable to remove this thickness of rock, and therefore a total volume of about 680 m³, 75/1.5 = 50 blast holes are necessary.

The distance between the holes, at the bottom, is 1.5 m, but the mouths of the holes, on the tunnel wall, are only 16/50 = 0.32 m apart.

Assuming that the stemming length is 1.5 m, mostly sufficient, and assuming that all the holes are loaded in the same way, that is, for 8.5 m, it happens that just beyond the stemming, where the loaded part begins, the holes are at a distance of just 0.5 m, and therefore, the distance between the charges, which at the bottom hole is 1.5 m, in the strip surrounding the tunnel, is about 1/3 of this value; it follows that, since the burden is always the same, the specific consumption is approximately three times greater.

This implies that there is too much explosive in this band or that it is too little at the bottom. Therefore, a good distribution of the explosive in the rock is not obtainable by loading all the holes in the same way.

The procedure of sizing the holes for ring blasts proposed by Esov (1970) is shown below, offering a simple solution to the problem. Other procedures, of course, can be followed, but that of Esov shows the principle well: since the discrepancy of the powder factor along the holes cannot be completely abolished, at least a tolerable limit must be placed on this discrepancy; with this criterion, a variation from 1 to 2 can be accepted, that is, it is accepted that in some areas, there is a double specific charge, but never more than double that of the bottom hole. The problem could be solved very well, theoretically, with complicated deck charges, but a very long time and demanding attention would be required to do that (charging the rings is in itself particularly slow).

The sizing according to the method proposed by Esov assumes that the spacing between the ends of the holes (the toe spacing) is equal to the burden; this means that the bottom hole pattern is square, as is quite common in ring blasting; the method is based on the calculation of the ratio of the burden to the diameter, that is, the ratio of the spacing to the diameter suitable for the explosive-rock pair considered.

The formula, transcribed with the original symbols used by the author, is:

$$w = \frac{0.427 \cdot b \cdot d \cdot Q}{f}$$

where w is the burden [cm], b is the in-the-hole charge density [g/cm³], d is the hole diameter [cm], Q is the specific heat of explosion [kcal/kg], and f is the Protodyakonov index.

The units of measurement are those used in the early 1970s; anyhow, the above-reported formula deserves a brief reflection: first of all, it should be noted that, despite its peculiar aspect, it is dimensionally correct: in fact, the constant 0.427 is simply the conversion constant from kcal/kg to kgm/g (1 kgm is equivalent to about 10 J); therefore, the product $0.427 \cdot b \cdot Q$ expresses the specific volumetric energy of the explosive in kg/cm³. This quantity, being a work/volume ratio, is physically a pressure. On the other hand, the Protodyakonov index corresponds to 1/100 of the uniaxial compressive strength of the rock expressed in the old unit (kg/cm²) and has, as well, the dimensions of pressure. Therefore, the ratio $0.427\frac{b \cdot Q}{f}$ is dimensionless.

The formula expresses the proportionality between the burden/diameter ratio (dimensionless) and the ratio of the specific volumetric energy of the explosive to the uniaxial compressive strength of the rock, which is also dimensionless.

In addition, if it is accepted that the specific energy is proportional to the explosion pressure (an acceptable assumption within a restricted class of explosives) and that the tensile strength of rocks is proportional to their uniaxial compressive strength (which is roughly true), the formula can be reduced to a simplified expression of the static sizing criterion: the ratio of the spacing to hole diameter is proportional to the ratio of the explosion pressure to rock tensile strength.

By adopting the S.I. units, as 1 kgm/cm³ means 10⁷ J/m³, that is, 10 MPa, and since a unit of the f Protodyakonov scale corresponds to 100 kg/cm², that is, $\cong$ 10 MPa, the formula simply becomes:

Spacing/diameter = [(explosive specific energy x explosive density)/rock's uniaxial compressive strength]: the specific energy is expressed in MJ/kg, the density in kg/m³, and the uniaxial compressive strength in MPa.

The formula has been recognized to be valid in very strong rocks (quartz diorites), blasted with nitrate explosives. However, it is not valid in general: in fact, it suggests redundant charges, but it is still conceptually interesting.

If the diameter, the explosive, and the rock are known, through the formula, the burden or the spacing can be calculated (that, with this type of sizing, are the same), and, knowing the perimeter of the ring thanks to the ratio of the perimeter to the spacing, the number of holes can also be obtained. Of course, the data could be different: for example, the spacing could be imposed and the diameter to be defined, or even it might be necessary to choose the appropriate explosive for a given blast when the geometry is already known. The charge is expected to be continuous, and the problem is to assign different charge lengths to different holes, to get a reasonably uniform charge distribution in the volume to be blasted.

The criterion adopted can be the following: once a hole (the longest) has been charged for the whole length minus the stemming (which is considered equal to the burden), the slice of rock, of width equal to the toe spacing (half on one side, half on the other), around this hole has the amount of explosive it needs: no other charge should enter this slice because that would cause an excess of charge. Therefore, the charge of the other holes must stop when more than half of the toe spacing approaches the charged part of the first hole: the limit is identified graphically by drawing two lines parallel to the hole, ½ of the toe spacing away from it. Once the length to be loaded for the two holes adjacent to the first (right and left of it) has been established, the same check must be carried out for the holes that are gradually farther away. An example is shown in Figure 7.6.

Obviously, placing a "respect slice" equal to half of the toe spacing around each charge does not ensure the perfect uniformity of the distribution, which is obtainable only in blasts with parallel holes. It is also worth mentioning that charging is not necessarily continuous in the case of very long holes, especially if the correct CPD must be respected; in this case, decking is employed with two or more in-the-hole delay times.

Figures 7.7–7.12 show some examples of underground production blasts. Description and details pertaining to figures are reported in captions.

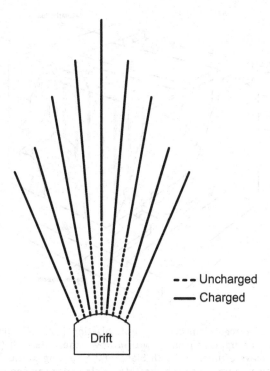

Figure 7.6 The interval charge design for ring-pattern blasting, with the indication of the common design used in the current ring blasting operation.

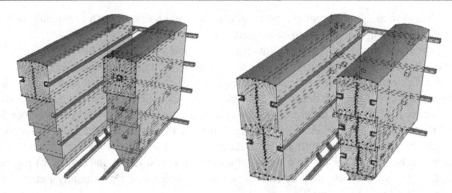

Figure 7.7 Left: exploitation with three sub-levels and one drift per sub-level (average specific drilling 0.22 m/m³; average powder factor 0.53 kg/m³; evolution of the scheme towards two drifts per sub-level (average specific drilling 0.31 m/m³; average powder factor 0.46 kg/m³). This involves an increase in development, but a better control of the walls. Right: evolution of the previous scenario (three sub-levels, two drifts per sub-level) towards two sub-levels and two drifts per sub-level. This scenario optimizes the specific drilling and leads to a reduction of developments. Each blast involves a volume of approximately 3,300 m³, and the average volume of a chamber (stope) is 500,000–550,000 m³ – width 30 m, length 180 m, and height 110 m.

Source: Cardu et al. (2016)

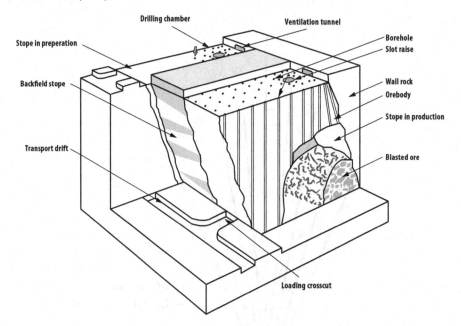

Figure 7.8 An example of a large-diameter long-hole blasting design. Drilling occurs throughout the stope at the upper chamber with a down-the-hole drill (Φ = 110 ~ 165 mm, four or five rows of holes in each blast, and a spacing of 2.0–2.4 m). The first few cutting blasting operations with the aid of a slot raise can generate more compensation space and a larger free surface. After the last layer of ore on the roof has been broken, the ore body that remains in the stope will be blasted via two or three lateral blasts. LHD = load, haul, dump machine.

Source: Wang et al. (2018)

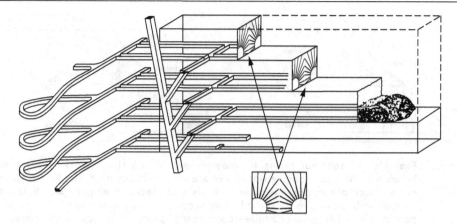

Figure 7.9 Longitudinal open-stope mining or sublevel retreat. Much of the development necessary for this mining method can be kept in the orebody. This is a relatively uncommon open-stope mining method in Canada but is very popular in Western Australian mines.

Source: Modified from Potvin and Hudyma (1989)

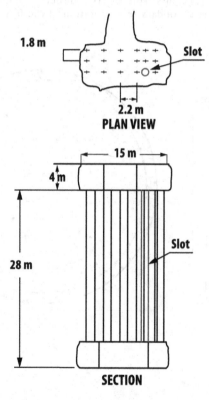

Figure 7.10 An example of a typical schematic primary stope drilling pattern, where a slot is developed by enlarging a 1.07 or 1.30 m diameter slot raise to the width of the stope, using parallel-hole blasting. The ore is fragmented in the stope using long parallel (primary stopes) or ring holes (secondary stopes), and mucked from a drift, oriented perpendicular to the stope strike, at the base of the stope. The top chamber is slashed (starting from a drift) up to the full stope strike length to allow drilling of 100 mm diameter parallel blast holes, typically with 2.5 m burden and 2.0 m spacing, with an off-center 1.07- or 1.30-m diameter raise-bore slot.

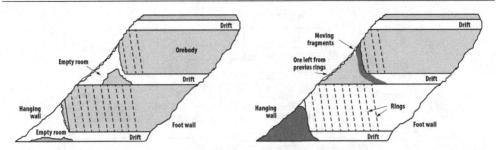

Figure 7.11 Examples of sublevel caving in underground. When the activity is performed beneath an inclined hanging wall, there are two methods used: the first is to load out as much ore as possible after each blast and leave an empty space below the hanging wall, as shown on the left. The second method is to extract only a small part of the ore fragments and leave the rest in place, as shown on the right. The blasting in the first method is similar to that of the first row in open-pit mines when the front face of the first row is completely free and nothing stands in front of it. On the contrary, in the second method, since the fragments from previous rings are left in front of the ring to be blasted, the boundary between the front face of the ring to be blasted and the remained fragments is partly a free surface. This makes the fragments from blasting directly collide with the remained fragments, making the secondary fragmentation of the fragments possible.

Source: Modified from Zhang (2017)

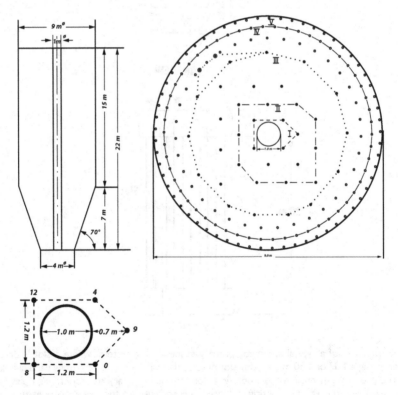

Figure 7.12 Construction of an underground silo by means of successive extensions, with concentric blasts of vertical holes, of a central raise.

Source: Adapted from Henniger and Borgmeyer (1977)

The exploitation proceeds, evidently, by means of horizontal descending benches, which recall the typical configurations of open-cast mining. Each sub-level is served by two drifts, from which ring blasts are made.

The silo, excavated in salt rock, was used for the storage of raw salt.

The central raise is obtained from a through-hole with a diameter of 60 mm, subsequently widened to a diameter of 1 m by dissolution.

In Figure 7.2, A: vertical section of the silo; B: a plan of the arrangement of the holes for the excavation of the upper cylindrical part, and of the widening to 4 m of the raise in the last 7 m (the conical part was subsequently excavated by slashing the raise with inclined holes); C: timing used in zone I.

In the excavation of the cylindrical part, the volume was divided into five concentric zones (I–V), each of which was excavated with a separate blast. The contour holes (zone V) and buffer holes (zone IV) are loaded only with a high-weight detonating cord (100 g/m), the others with ANFO bulk, and a high-weight detonating cord along the hole. Hole diameter: 60 mm.

Timing: I–III blasts: micro-delays; blast IV: only two delays (internal and external crown); note the two breaking blast holes that start with blast III to avoid the arching effect; blast V: instantaneous. The average specific consumption is 300 g/m³.

7.2 Blasts for blind-bottom excavation

In these types of excavations (tunnels, shafts, and raises), the maximum specific consumption is recorded, both explosives (systematically higher than 1 kg/m³, with peaks over 10 kg/m³), drilling (systematically higher than 1 m/m³, with peaks over 15 m/m³), and detonators (systematically higher than 0.3 pcs/m³, with peaks over 15 pcs/m³), together with the maximum complexity of the blast, which always includes at least two, more often three, functional groups of holes, that is, groups of holes which, having a different function, are sized according to different rules.

Indeed, of the main two or three functional groups (opening, production, eventual contour holes – see Figure 7.13), only the first has new characteristics and is sized according to criteria that were not examined in the discussion of the other types of blasts.

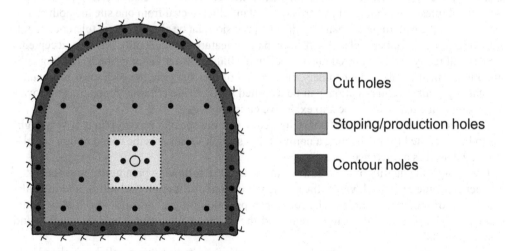

Cut holes

Stoping/production holes

Contour holes

Figure 7.13 The blast design concept for tunneling applications.

The group of production holes is made by a blast with parallel or sub-parallel holes, similar to that of bench blasting and, besides the different (much greater, with the same explosive/rock pair) specific consumption assigned to them, can be treated the same way.

The group of contour holes has the main function of guiding the fracture and the detachment, similar to what occurs in trenches: however, in the excavation of tunnels, the smoothing solution (where the contour holes are triggered last) is universally preferred to the pre-splitting. An important difference between tunneling contour holes and those for trenches or profiling walls is that in the former case, they cannot be exactly parallel to each other and to the production holes but must diverge from the excavation axis by a small angle (look-out) to ensure section preservation in subsequent blasts.

The topics treated in this chapter do not necessarily cover the blind-bottom blasts but provide at least a schematic framework. These topics are

1   the average specific consumption;
2   the rules for placing and triggering the cut-holes;
3   the calculation of the charges for inclined cut-holes;
4   the calculation of the charges for parallel cut-holes;
5   shafts and raises blasting.

7.2.1 Drill and blast in tunneling

Drill and blast is a common technique for tunnel driving, either for mining or for civil purposes, at least for tunnel lengths lower than 3–4 km: it is a mature technology, offering a huge variety of cases to analyze.

As the name of the technique indicates, three factors are involved: explosive, rock, and drilling. The main goal of tunnel excavation is to assess the relative importance of each of them and to describe the techniques commonly employed to get a satisfactory result.

Several rounds in drifting operations, covering a variety of rock-drilling system-explosive varieties can be surveyed; the rock can exhibit important differences from one site to another; drilling can be performed in the range "manually steered jumbo" – "computerized jumbo", but the general rules for planning a blast are not substantially affected from one site to another. The sensitivity of the pattern performance to drilling precision can be also analyzed; tolerance standards for drilling machinery related to drilling pattern features are generally, anyway, accepted.

Some difficulty can be inferred due to specific drilling, powder factor, or both. Drilling consumption, which involves the capillary distribution of the charge in the rock mass to obtain a satisfactory grain size distribution, can be assumed as a reliable criterion based on a narrower dispersion of the data with respect to explosive consumption.

The effect of the cross section (whose influence on both specific consumptions is very high) has to be accounted for by defining a normal cross section versus specific consumption correlation, to be used as a reference line.

To establish the quality of the results of the blasts, the following parameters must be taken into account: the overbreak value, the efficiency (actual pull/design pull ratio), and the half-cast factor of the contour holes. The relationships linking these indicators of quality to the features of the rock, the drilling system, and the charge must be analyzed (Kecojevic and Radomsky, 2005).

The cross sections can vary from a few m^3 to up to $160\ m^3$, with either parallel or inclined hole cuts and a wide range of explosives and initiation systems can be used.

The result of a blast is a volume V of broken rock equal to the product of the excavation cross section S by the pull l.

For this purpose, a certain total amount of explosive Q is distributed, according to specific rules, in volume V.

The average specific explosive consumption of the blast, P.F., is the ratio, expressed in kg/m³, of Q to V, that is:

$$P.F. = \frac{Q}{S \cdot l}$$

The values of S and l are project dependent. On the final balance (blast performed), different powder factors can be obtained, as the cross section, the pull or both could be different from those expected (an infinite value could also be found if the blast fails completely).

S has a value that the blast performer cannot modify. l must be chosen on the basis of the capability and size of the drilling machines, the type of blast chosen, the characteristics of the clearing system, the need or not for static support interventions during the excavation, or any limitations to total employable charges, the general organization of the work cycle (i.e., of the times devoted to the various phases of it), and basically, it increases as the cross section increases. The range is quite extensive, from 1 m to 5 m, with peaks, in exceptionally favorable or experimental cases, of over 7 m.

To distribute the charges in volume V, it is obviously necessary to drill a certain number of holes, according to requirements that ensure good efficiency for the charges. The drilling diameters used vary within a rather narrow range, typically between 30 mm and 50 mm. The total length of holes L needed for a blast tends to grow as Q increases, as the overall volume of holes must always be greater than the explosive volume, but there is not a very close dependence between L and Q, as the holes can have different diameters, the explosives different densities, and the utilization coefficient (ratio of the charged length to the total length) of the holes, as well as the coupling ratio of the charge to the hole (ratio of the diameter of the charge to the of the hole), can vary.

As concerning the specific drilling, drill and blast in tunneling can be expressed as: the ratio of L to the volume that the blast must remove. Consequently, it is expressed in m⁻² or in m/m³:

$$S.D. = \frac{L}{S \cdot l}$$

Finally, the blast consists of a certain number of blast holes, and each of them requires at least one detonator. The initiation system with detonating cord is usually avoided in tunneling, as it has various drawbacks (negative oxygen balance following the decomposition of PETN, and risks of interruption of the circuit due to the possible overlapping of strands of cord, as the detonation, which occurs at a speed of about 7,000 m/s, induces a violent shock wave). In Italy, both the electric system and the Nonel are commonly used, while the electronic detonators are widespread abroad, at least for the last 25 years, becoming very popular, especially in tunneling.

The number of blast holes, in a blast with parallel-hole cuts, is given by the ratio of the drilled length to the theoretical pull; in blasts including inclined holes, it may differ, but not too much, from this ratio.

The specific consumption of detonators D.C. is the ratio of the holes' number n to the blasted volume, and is expressed in the number of *pieces/m³*:

$$D.C. = \frac{n}{S \cdot l}$$

On the basis of the previous statements, it can be said that the main specific consumptions of a tunnel excavation can be roughly correlated to one of them, that is, the powder factor, as follows:

- The specific drilling is approximately proportional to the powder factor.
- The specific consumption of detonators is approximately proportional to the ratio of the specific drilling to the pull.

It is useful to have available a tool to predict, even approximately, the specific consumption to be expected for a given excavation: this roughly allows to have an idea of the other consumptions as well.

The powder factor depends on various factors, among which the main are listed in hierarchical order, that is, from the most influential to the least influential:

1) the cross section;
2) the explosive-rock pair;
3) the type of blast.

Such order is established based on the analysis of several cases (Mancini et al., 1995a, 1995b), but it has logical reasons too.

The cross section is the most important factor, as it has a double effect: as the cross section decreases, the percentage incidence of blast holes operating in difficult conditions increases (with few free surfaces, with limitations to the opening of the crater or of the detachment wedge, with the need to break the rock by shear rather than by tensile stress), and in return, as the cross section decreases, the need of obtaining a finely fragmented material increases (Onederra et al., 2004), to allow it to be quickly removed with the equipment suitable for a small construction site. Furthermore, in very small cross sections, exceeding the specific charge to achieve greater certainty of effect does not cause serious economic damage, with the same explosive-rock pair and the same type of blast.

The powder factor can vary by a whole order of magnitude, from the smallest sections (access tunnels) to the largest (motorway tunnels).

The explosive-rock pair apparently has a much smaller influence on specific consumption: "apparently" because, in practice, more powerful (and more expensive) explosives are used in stronger rocks. With the same section and type of blast, changing the explosive-rock pair can still lead to a difference in the powder factor from 1 to 2 or a little more, judging by the case studies. The type of blast has apparently an even less influence than the two other factors. Also, in this case, it is said "apparently", as the type of blast is also chosen based on the cross section and the explosive-rock pair, and the case histories analyzed show only the choices that have given satisfactory results. However, there is a tendency for higher powder factors, all things being equal, in blasts with parallel-hole cuts (this negative trend is, of course, offset by other advantages).

Mancini and Pelizza (1969), based on a statistical analysis of an abundant series of excavations of civil and mining tunnels, have proposed a correlation formula for the prediction of the powder factor:

$$P.F. \cong \left(\frac{10}{S} + 0.6\right) \cdot A \cdot B \cdot C$$

where A, B, and C are coefficients that consider the type of rock, explosive, and blast (in particular, the type of opening).

The rocks are classified into five classes corresponding, in principle, to the first five of the Protodyakonov classification (0–IV); it should be noted that, for the latter, the use of explosives is exceptional, since mechanical excavation is better.

The values of coefficient A (Rock coefficient) are shown in Table 7.1.

Explosives are classified into four classes, with decreasing power and density, from the most powerful and heaviest (straight dynamites) to the least powerful and light (nitrate-based explosives). The latter, although sensitive to humidity, are rarely and almost never used to load cut holes and underneath holes. However, the progressive decrease in the consumption of dynamite over the years has been characterized by the progressive introduction of explosive emulsions, both in cartridges and loose, with increasingly frequent applications underground. The values of the B coefficient are shown in Table 7.2.

The density referred to is that in the cartridge. The density in the hole, especially when charging with pre-packaged charges (a fairly common practice), is significantly lower. The most used explosives are those of classes 2 and 3.

The types of blasts are classified into three classes: opening with inclined holes (pyramid, wedge), fan-holes, and parallel holes. The second type is less common and is used almost exclusively in stratified rocks, orienting the fan to take advantage of the joints for the expulsion of the rock. The third is generally used in small cross sections but is also applicable to bigger sections.

The values of the C coefficient are shown in Table 7.3.

The correlation formula is intended as a rough forecasting tool to find the powder factor and evaluate the total amount of explosive required by a blast.

Table 7.1. Rock coefficients.

Class	Protodyakonov class	Examples	Coefficient
1	0	Quartzite, compact porphyry	1.3
2	I	Sound granite and gneiss	1
3	II	Compact limestone	0.9
4	III	Phyllite and clay-schist	0.8
5	IV	Porous limestone, marl, gypsum	0.5

Table 7.2. Explosive coefficients.

Class	Explosive type	Coefficient
1	Straight dynamites ($\gamma > 1.5$)	0.95
2	Dynamites with a high NG content ($\gamma > 1.4$)	1
3	Other explosives with $\gamma > 1.2$	1.1
4	Nitrate-based explosives ($\gamma < 1.2$)	1.2

Table 7.3. Blast coefficients.

Class	Type of blast	Coefficient
1	Inclined holes (pyramid, wedge)	1
2	Fan holes	0.9
3	Parallel holes	1.45

With a little common sense, it can also be used as a tool for designing a blast, if the term "design" is understood as adapting a blast plan to a specific situation that has proved satisfactory in other cases.

In such cases, it is sufficient

- to design and compute (amount of charge and delay times), a suitable cut-hole scheme, as clarified below with some examples;
- to fill the remaining part of the section with a suitable number of production holes.

As for specific drilling, in the range of the most commonly used diameters (30–35 mm), the correlation formula is expressed by:

$$S.D. \cong 2.3 \cdot \left(\frac{10}{S} + 0.6 \right) \cdot A \cdot B\text{'}$$

where $B\text{'}$ is a coefficient taking into account the type of explosive, as shown in Table 7.4.

With diameters greater than 35 mm, the specific drilling can, of course, decrease. However, when prepacked charges are adopted, the influence of the drilling diameter on the specific consumption of holes is negligible.

The formula tends to provide conservative estimates in the case of blasts which, for reasons of accurate profiling, resort to accurate smoothing (due to the thickening of the contour holes).

7.2.2 Rules for designing and timing the cut-holes

The pull of a blast depends on the pull of the cut, which can be considered a small blind cut that will be enlarged, a few fractions of a second later, by the production holes.

The case of inclined holes' cuts must be considered separately from that of parallel-holes' cuts; in the former, in fact, the pull is conditioned by the cross section, whereas in the latter it is independent of it (but it is, of course, also limited by other factors).

In fact, the pull ideally corresponds to the maximum distance (in the forward direction) from the face that can be reached by the longest cut-hole: if the latter is inclined with respect to the face, this distance is limited by the need to avoid interference between the drilling system (drilling rig and/or operator) and the tunnel wall.

Before drawing a possible arrangement of the cut-holes on the assigned cross section, it is, therefore, necessary to check (with simple geometric constructions) if they can be drilled with the available drilling machine.

The example that is given in Figure 7.14 refers to the holes to be drilled for a V-shaped opening, in a central position, with a drilling rig mounted on a 4-m-long guide. The objective is to obtain a 60° sharpness of the wedge to be ejected, a value compatible with the success of

Table 7.4. Explosive coefficient $B\text{'}$ for computing the specific consumption of blast holes.

Class of explosive	Coefficient $B\text{'}$
1	0.6
2	0.65
3	0.80
4	1.2

this type of opening. In the case of a 6-m-wide tunnel (case 1), the maximum pull that can be obtained respecting the imposed geometry is 1.2 m; in the case of a 10-m-wide tunnel (case 2), it reaches 4.7 m. Of course, these are neither the only-possible solutions nor the best.

In case 1, to obtain a greater pull, for example, 3 m, an asymmetrical fan opening could be used, like the one shown in the same Figure 7.14, at the bottom left.

In case 2, a 5.2 m pull would have been considered overdone: a 4 m or lower pull would have been satisfactory. Furthermore, holes would have been added to help the fragmentation (baby cut), to avoid the risk of detaching too large blocks from the central part of the opening, as shown in the same Figure 7.14, bottom right.

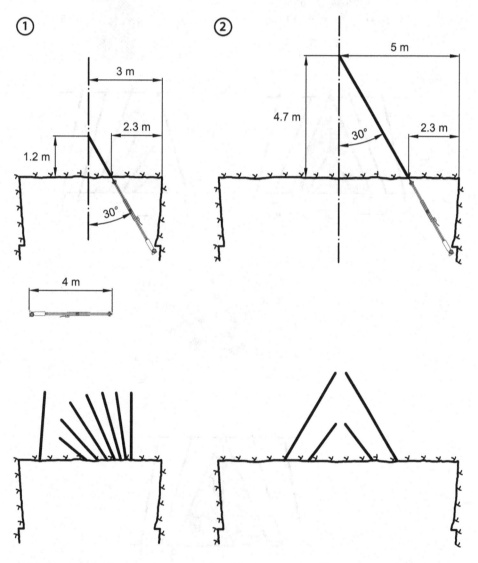

Figure 7.14 Geometrical constraints to inclined holes round drilling and possible adaptations for narrow tunnels.

An attempt has to be made to assign an opening of 60°, or greater, to the central wedge. If this does not allow an adequate pull to be achieved, sharper wedges can be drilled (lower than 60°), but the powder factor must be increased to obtain a satisfactory ejection; otherwise, it is better to switch to another type of cut in general.

Figure 7.15 shows some examples of cuts with inclined holes.

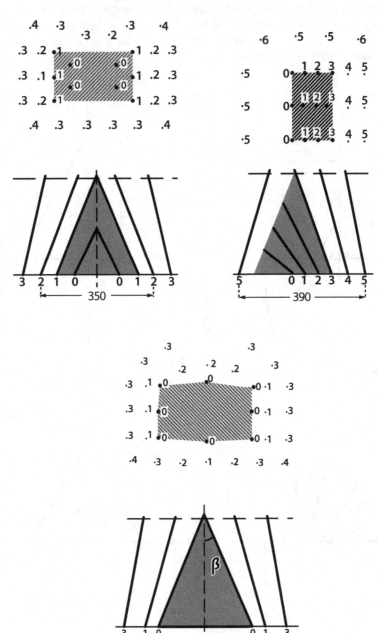

Figure 7.15 Examples of openings with inclined holes.

As for timing, micro-delays (20–30 ms) are generally used for cut holes and ordinary delays (1/4 s, ½ s) for production holes (Berta, 1990).

When parallel-hole cuts are realized, the first detonating holes have as a free surface the walls of one or more dummy holes parallel to them, drilled at a short distance. Some rules should be followed to warrant the correct operation, yet many schemes in use do not respect them, despite working properly.

As an example, the simple spiral opening is considered, with four charged holes and a central dummy hole, shown in Figure 7.16: each hole, at the instant of the explosion, should see the free surface under an angle large enough to allow the smooth detachment of the rock segment

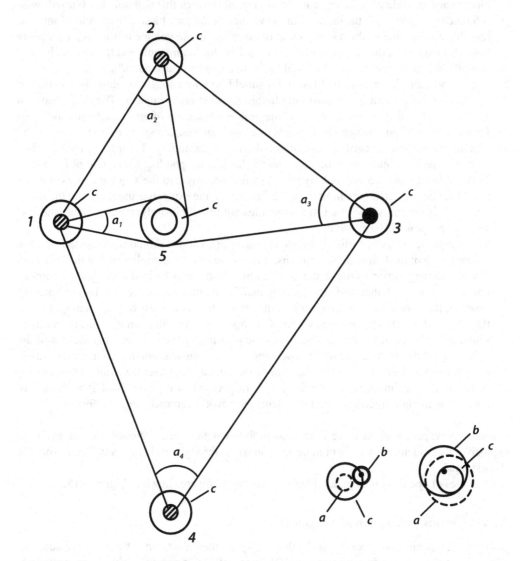

Figure 7.16 Example of a spiral opening. 1–4: charged holes; 5: dummy hole; c: circles of uncertainty. Meaning of the uncertainty circle: a: project position; b: possible actual position; c: circles of uncertainty.

between the charged hole and the free wall. With reference to Figure 7.16, hole #1 sees the free surface (the dummy hole) under angle α_1, hole #2 sees its free surface (the opening created by hole #1) under angle α_2, and so on.

As for the minimum value to be assigned for angle α_1, different authors suggest different limits (Singh and Xavier, 2005; Zare and Bruland, 2006), included in the range 20°–30°.

The following five points summarize the main rules to follow when performing parallel-hole blasts.

1) At the time of the explosion of each blast hole, the empty volume available must be suffi-cient to accommodate the increase in the volume of the rock that that hole has blasted. With reference to Figure 7.17, the increase in the volume of the rock blasted from hole #1 must be less than the volume of the dummy hole; the increase in the volume of the rock blasted by hole #2 must be less than the volume represented by the dummy hole and the volume blasted from the first hole, and so on. The swell factor to consider is around 50%.

2) The delay sequence among the blast holes should be long enough to allow the ejection of the rock on behalf of a given blast hole before the next one detonates. The delay between explosions should therefore increase as the pull increases. An often-quoted value is 20 ms for each meter of pull, which corresponds to an average ejection speed of 50 m/s.

3) The minimum distance between two blast holes (in the example of Figure 7.17 the two clos-est are #1 and #2) must never be less than the limit corresponding to the risk of flashover: in this case, the two holes would detonate simultaneously and the sequence of explosions would not be respected. The limit distance depends on the explosive, the rock, and the linear charge: it is generally 20–30 cm for dynamites with high nitro-glycerine content, and it is less for less powerful explosives.

4) The geometric conditions (distances, angles) mentioned above should be respected over the whole length of the holes: at their mouths, it is easy to enforce compliance, but the holes can deviate, so the position of the centers of the holes' bottoms is defined only by the circles of uncertainty in which they may accidentally match, and they are larger with longer holes. In practice, the condition is reduced to a limitation of the slenderness of the opening, that is, the ratio of the pull to the average width of the opening. A practical limit of this ratio can be found in the range of 8–12 as a function of the positioning accuracy and parallelism of the holes (e.g., if the average width of the opening is 40 cm, the maximum recommended pull is just over 3 m in the first case and about 5 m in the second; if the average width of the opening is 50 cm, the maximum recommended pull is respectively 4 m and 6 m). Higher ratios are possible with high-precision guiding systems (Sandvik – Tamrock Corp., 1999).

A blast that respected all of these limits has probably never been designed, but, in any way, explosives that did not respect them in design can work, through mechanisms different from the one assumed.

Examples of openings with parallel holes of common use are shown in Figure 7.18.

7.2.2.1 V-cuts: calculation of the charges

Only wedge openings are examined (V, double V), as they are the most commonly adopted, with respect to the procedure proposed by Olofsson (1991): it considers simple or multiple V openings (the example in Figure 7.18 is a double V opening) on a number of rows suitable to cover the desired cutting height C (usually 3 rows, a solution to be considered standard) and a

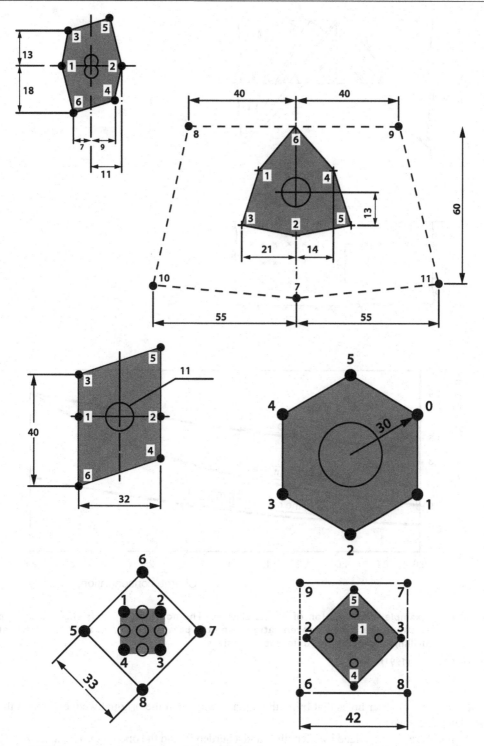

Figure 7.17 Examples of openings with parallel holes.

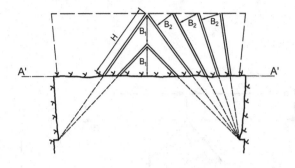

A'-A' section

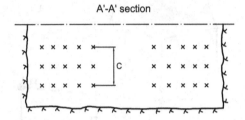

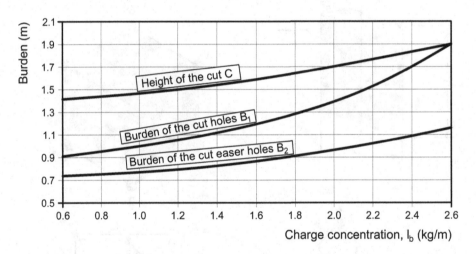

Figure 7.18 Example of calculation of V-cuts charges. The nomogram shows the relationship between the charge concentration l_b and the parameters B_1, B_2, and C for different drilling diameters and different explosives.

Source: Olofsson (1991)

certain number of easer holes that bring the initial cavity from the original wedge shape to the parallelepiped shape.

The holes are characterized by a length H and a burden B, and the opening is characterized by a height C, as shown in Figure 7.19.

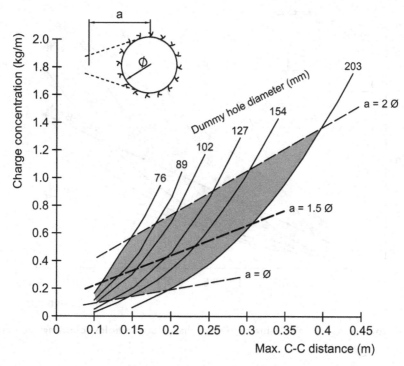

Figure 7.19 Nomogram for calculating the linear charge of the first blast hole.

Source: Olofsson (1991)

To calculate the opening, the bottom charge (charge concentration, l_b) is first calculated, which depends on the drilling diameter, the diameter of the cartridges, the density of the explosive in the cartridges, and the more or less complete compaction of the explosive in the hole. Olofsson provides indicative l_b values for four commercial explosives, and for a range of diameters between 30 and 51 mm; if different values are used, the calculations must be adapted to other schemes.

Once l_b is known, B_1, B_2, and C, which are necessary to draw the blast, are found through the nomogram in Figure 7.19 (the acuity of the most advanced V is set equal to 60°). Of course, the geometric compatibility with the cross section must be checked separately.

Otherwise, if the desired pull is known, B_1 can be calculated, assuming the most appropriate number of wedges (from 1 to 3, considering the need to distribute the explosive well in the volume to be expelled), identifying the suitable values of l_b and, consequently, the diameter and the height C.

The holes of the cut are loaded, from the half bottom of the hole, with a linear charge l_b and, to the remaining part, with a reduced linear charge (30%–50% l_b), leaving a stemming length equal to 0.3 B_1 for the holes of the V-cut and 0.5 B_2 for the easier holes.

7.2.2.2   *Parallel cut-holes: calculation of the charges*

For the calculation of the linear charge (kg/m) of the first blast hole, which has only the dummy hole as a free surface, the nomogram shown in Figure 7.20 can be used, according to Olofsson (1991).

The other holes benefit from a greater free surface and the relative linear charge can be calculated from the nomogram in Figure 7.21, by Olofsson (1991).

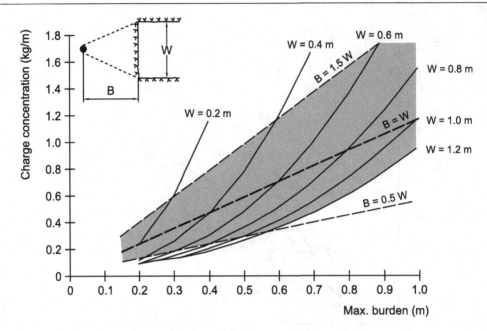

Figure 7.20 Nomogram for calculating the linear charge of the holes following the first one.
Source: Olofsson (1991)

However, it should be noted that many types of openings are not suitable for this scheme. An empirical procedure, in any case valid, can be based on the correlation between the powder factor and the cross section, which was already analyzed for the evaluation of the average specific consumption: it is also valid for the calculation of the average specific consumption of the parallel-hole cuts, which can be considered a small tunnel. An opening with a slenderness compatible to the wanted pull is chosen, its section S_c is established, included in the usual range of parallel-hole cuts, and the specific consumption pertaining to the value of S_c is calculated.

Multiplying this specific consumption (generally a few tens of kg/m³) by the product $S_c * l$, the total charge to be assigned to the opening, in kg, can be obtained; it will be distributed between the four and eight holes pertaining to the cut.

7.2.3 Examples of blasts for tunnel excavation

Figures 7.21 and 7.22 have been chosen for some detail worthy of reflection.

7.3 Blasts for shaft sinking

When the shafts are excavated from top to bottom, the pull is limited by reasons of difficulty in mucking the material, and therefore, cuts with parallel holes are of little interest.

Openings with inclined holes are almost always used (some examples are shown in Figure 7.23).

The specific consumption of explosives and holes is increased by 30%–50% compared to those of a tunnel with the same cross section and explosive-rock pair, as the material has to be very well fragmented, in order to be suitable for collection by a clamshell.

Particular solutions can be imposed by the flow of water: it settles on the bottom of the excavation and is removed with pumps, drawing it from the lowest point. But, if the bottom is flat

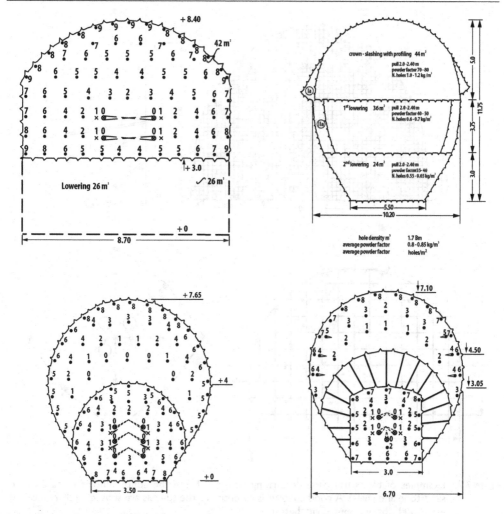

Figure 7.21 Examples of excavation of large cross-section hydraulic tunnels.

A: excavation carried out advancing in the crown (central V opening, pull 1.95 m, specific consumption 1.18 kg/m³) on the 42 m² cross section, and subsequent lowering to increase the cross section by a further 26 m², with blasts of horizontal or vertical holes. The average specific consumption (progress in crown + decline) is 0.9–1 kg/m³.

B: splitting of the cross section for the excavation of a 104 m² tunnel. The phases are advancement in the crown over a section of 44 m² (slashing with profiling – Ia area), 2–2.4 m pull, specific consumption 1–1.2 kg/m³; lining of the crown and first lowering with 36 m² cross section, first respecting the walls IIa which are then profiled (blasts of horizontal holes, pulls 2–2.4 m, specific consumption 0.6–0.7 kg/m³); second lowering with 24 m² cross section (blasts of horizontal holes, pulls 2–2.4 m, specific consumption 0.55–0.65 kg/m³) and completion of the lining. Average specific consumption over the whole section: 0.8–0.85 kg/m³.

C: advancement with a base tunnel (16.3 m², V-cut, pull 2.7 m, specific consumption 1.55 kg/m³) and enlargement with blasts of horizontal holes, (section 27.4 m², pull 2.2 m, specific consumption 0.9 kg/m³), to reach the final 43.7 m² cross section.

D: advancement with a base tunnel (9 m², V-cut, pull 1.57 m, specific consumption 2.05 kg/m³) and two-stage widening: slashing with a semi-ring (section 7 m², specific consumption 0.7–0.8 kg/m³) and second enlargement with horizontal blast holes (section 20 m², pull 3 m, specific consumption 0.9–1 kg/m³), up to the final cross section of 36 m². The average specific consumption is 1.17 kg/m³.

Source: Pulg (1972)

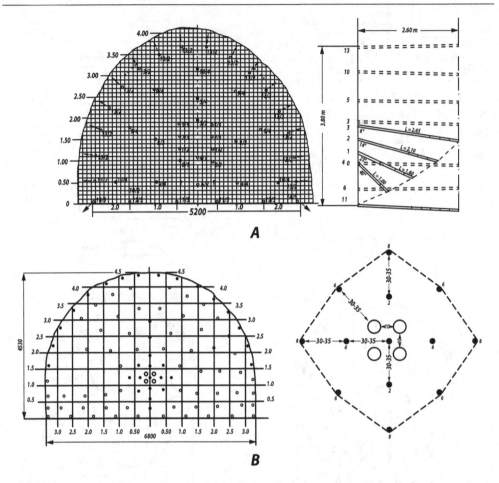

Figure 7.22 Examples of blasts using the detonating cord to charge the contour holes (excavation in sandstone and shale). A good contour is essential, as the tunnels are subsequently reinforced with steel ribs, without spritz beton.

> A: fan cut blast, section 18.4 m², pull 2.6 m; contour holes (crown) loaded with two dynamite cartridges (250 g) and 40 g/m detonating cord (Jankowiak and Von Depka, 1974).
> B: blast with parallel holes cut, section 28.1 m², pull 3 m; contour holes (crown) loaded with two cartridges of dynamite and 40 g/m detonating cord (Erlaker, 1977).

and the tributary flow is large, its level can grow all over the cross section of the shaft, making the work impossible or very demanding. Therefore, an alternative can be to proceed with a partial section, proceeding consecutively in the right and left half of the section, usually with fan blasts, or with V-cuts, to always have half of the section dry for drilling (Figure 7.24).

7.3.1   Blasts for raising

The raises are generally realized from bottom to top, with blasts like those used for tunnels, almost always with parallel holes (sometimes fan-shaped), as the narrow cross section does not allow satisfactory pulls with other types of blasts.

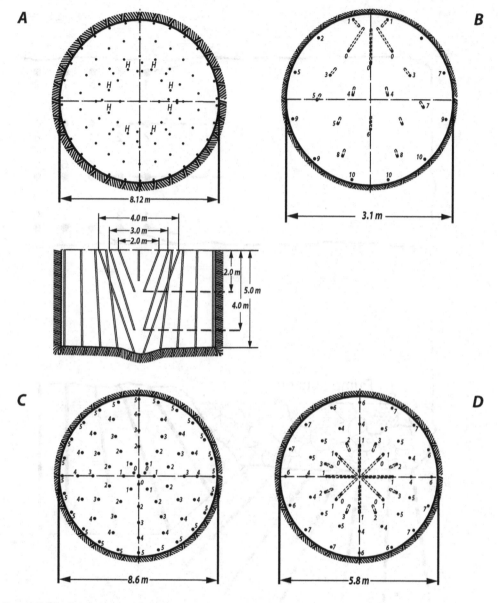

Figure 7.23 Conventional blasts

- A: plan and vertical view of blasts with a central multiple pyramid cut. It can reach high pulls, but it is practicable only with large cross sections.
- B: blast with an eccentric pyramid cut, for small cross sections.
- C: blast with parallel-hole cut, similar to those used in tunneling.
- D: Delay times adopted for a blast with multiple pyramid cuts (micro-delayed detonators).

Source: Wild (1973)

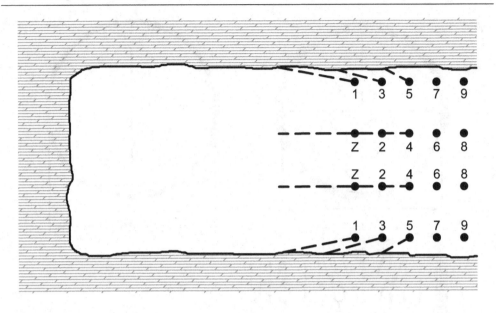

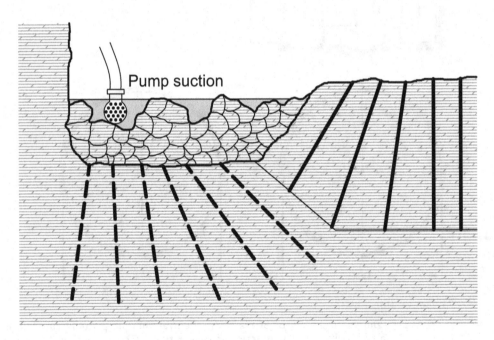

Figure 7.24 Blasts expressly designed for the excavation in the presence of water: plan view and vertical section of a split blast (fan cut), for the excavation of a rectangular raise.

If the excavation from bottom to top offers the convenience of removal, by gravity, of the muck, drilling is more difficult. For this reason, many short raises (a few tens of meters) are performed with the so-called long-hole system: the holes are drilled from the top, covering the whole length of the raise, then they are loaded, always from the top, and exploded in slices, from bottom to top, a few meters in length. From what has been said on slenderness, to implement this method, a special drilling system is required: the drill rods and, consequently, the holes, must have a diameter two to four times greater than those used in tunneling, and the drill guidance must be very rigid and accurate (the slenderness can even exceed 40). However, the drilling precision can be checked by examining the area, at the base of the future raise, into which they lead, and possibly re-drilling those that are too deviated.

Two types of blasts are commonly used in the long-hole method: conventional parallel holes with a dummy central hole, similar to those used in tunneling, and crater blasts. In this case, the holes are large in diameter and loaded with spherical charges ($L \leq 6\ \Phi$), which subsequently open craters downwards until the whole cross section is covered; it is an application of the V.C.R.

Examples are shown in Figures 7.25 and 7.26.

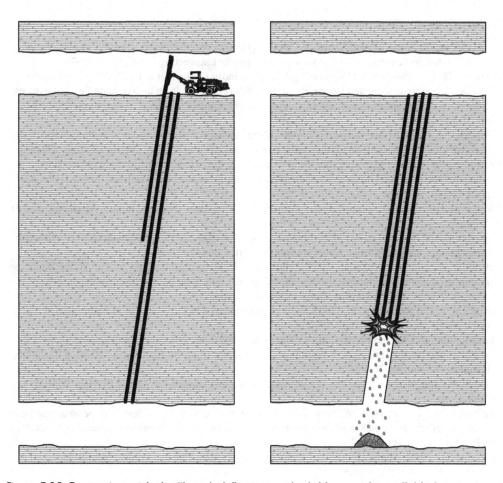

Figure 7.25 Excavation with the "long-hole" raise method, blasts with parallel holes.

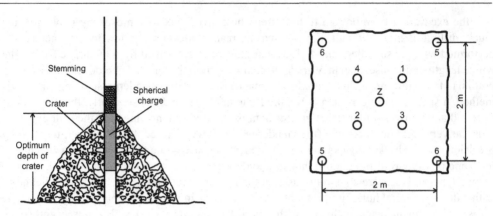

Figure 7.26 Left: VCR borehole performance; right: typical VCR pattern and delay sequence.

References

Ak, H., Iphar, M., Yavuz, M. and Konuk, A., 2009. Evaluation of ground vibration effect of blasting operations in a magnesite mine. Soil Dynamics and Earthquake Engineering, 29(4), pp. 669–676.

Berta, G., 1990. Explosives: An Engineering Tool. Italesplosivi, Milano. Ed. La Moderna, Novara, Italy.

Cardu, M., Dipietromaria, S. and Oreste, P., 2016. Sub-level stoping in an underground limestone quarry: An analysis of the state of the stress in an evolutionary scenario. Archives of Mining Sciences, 61(1), pp. 199–216.

Crackel Jr., M., Heisel, M. and Ramos, G.G., 1981. Stope blasting design and experience at the Carr Fork Mine. Proceedings of the International Conference on Caving and Sublevel Stoping, Denver, USA, pp. 529–533.

Erlaker, F., 1977. Beschleunigung der Bohr – und Sprengarbeiten beim Streckenvortrieb durch Verwendung von Patronen Größeren Durchmessers auf der Schachtanlage Westfalen. Nobel Hefte, 43(4), pp. 145–149.

Esov, T.S., 1970. Gornyi Zhurnal, 146, 12. Quoted in Ring Hole Blasting, Mining Magazine, 124(4), p. 335.

Henniger, G. and Borgmeyer, W., 1977. Abteufen eines Rohsalzbunker – Gesemts durch Sprengarbeit mit Sprengschnur Supercord 100 t. Nobel Hefte, 44(4), pp. 121–127.

Jankowiak, H. and Von Depka, H., 1974. Erfahrungen in Steinkohlenbergbau bei der Verwendung von Patronen Größeren Durchmessers (Gesteinsprengstoff) und beim Schonenden Sprengen mit 40 g – Sprengschnur in Streckenvortrieb. Nobel Hefte, 40(1), pp. 9–19.

Kecojevic, V. and Radomsky, M., 2005. Flyrock phenomena and area security in blasting-related accidents. Safety Science, 43(9), pp. 739–750.

Mancini, R., Cardu, M. and Fornaro, M., 1995a. An analysis of the influence of rock quality on the results of controlled blasting practices. Proceedings of the II Int. Conf. on Engineering Blasting Technique, Kunming (China), pp. 264–267.

Mancini, R., Gaj, F. and Cardu, M., 1995b. Atlas of Blasting Rounds for Tunnel Driving. Politeko Ed., Torino.

Mancini, R. and Pelizza, S., 1969. Previsione dei consumi di esplosivo e di lavoro di perforazione nello scavo di gallerie. Proceedings of the I International Conference on Technical Problems in Tunnelling, S.P.E. Ed., Torino, pp. 1075–1087.

Olofsson, S.O., 1991. Applied Explosives Technology for Construction and Mining. Nora Boktryckeri AB, APPLEX Pub., ÄRLA, Sweden.

Onederra, I. and Chitombo, G., 2007. Design methodology for underground ring blasting. Mining Technology, 116(4), pp. 180–195.

Onederra, I., Esen, S. and Jankovic, A., 2004. Estimation of fines generated by blasting – applications for the mining and quarrying industries. Mining Technology, 113(4), pp. 237–247.

Potvin, Y. and Hudyma, M.R., 1989. Open Stope Mining Practices in Canada, Presented at the 91st CIM Annual General Meeting, Quebec City, May.

Pulg, W., 1972. Ausbrucharbeiten bei der Herstellung des Pumpspeicherwerkes Vianden. Nobel Hefte, 38(1), pp. 31–40.

Sandvik – Tamrock Corp., 1999. Rock Excavation Handbook for Civil Engineering. Matti Heiniö Ed., Tampere, Finland.

Singh, S.P. and Xavier, P., 2005. Causes, impact and control of overbreak in underground excavations. Tunnelling and Underground Space Technology, 20(1), pp. 63–71.

Soltani-Mohammadi, S., Amnieh, H.B. and Bahadori, M., 2012. Investigating ground vibration to calculate the permissible charge weight for blasting operations of Gotvand-Olya dam underground structures. Archives of Mining Sciences, 57(3), pp. 687–697.

Wang, M., Shi, X., Zhou, J. and Qiu, X., 2018. Multi-planar detection optimization algorithm for the interval charging structure of large-diameter longhole blasting design based on rock fragmentation aspects. Engineering Optimization, pp. 1–15. ISSN: 0305-215X (Print) 1029-0273 (Online) Journal homepage: www.tandfonline.com/loi/geno20.

Wild, H.W. 1973. Der Sprengarbeit beim Schachtabteufen. Nobel Hefte, 39(4), pp. 122–133.

Zare, S. and Bruland, A., 2006. Comparison of tunnel blast design models. Tunnelling and Underground Space Technology, 21(5), pp. 533–541.

Zhang, Z.X, 2017. Kinetic energy and its applications in mining engineering. International Journal of Mining Science and Technology, 27(2), pp. 237–244.

Zhang, Z.X. and Naarttijärvi, T., 2005. Reducing ground vibrations caused by underground blasts in LKAB Malmberget mine. Fragblast, 9(2), pp. 61–78.

Index

Printed in the United States
by Baker & Taylor Publisher Services